COAGULATION AND ULTRAFILTRATION IN SEAWATER REVERSE OSMOSIS PRETREATMENT

Cover Photo: Satellite image acquired by Envisat's MERIS instrument on November 22, 2008 of the red tide bloom in the Gulf of Oman spreading to the Persian Gulf (Credits: C-wams project, Planetek Hellas/ESA).

COAGULATION AND ULTRAFILTRATION IN SEAWATER REVERSE OSMOSIS PRETREATMENT

DISSERTATION

Submitted in fulfilment of the requirements of
the Board for Doctorates of Delft University of Technology
and
of the Academic Board of the UNESCO-IHE
Institute for Water Education
for
the Degree of DOCTOR
to be defended in public on
Tuesday, 20 May 2014, 12:30 PM
in Delft, The Netherlands

by

S. Assiyeh ALIZADEH TABATABAI

Master of Science in Water Supply Engineering
UNESCO-IHE, Institute for Water Education

born in Tehran, Iran

This dissertation has been approved by the promotor:
Prof. dr. M.D. Kennedy

Composition of Doctoral Committee:

Chairman	Rector Magnificus, Delft University of Technology
ViceChairman	Rector UNESCO-IHE
Prof. dr. M.D. Kennedy	UNESCO-IHE/Delft University of Technology, promotor
Em. Prof. dr. ir. J.C. Schippers	UNESCO-IHE/ Wageningen University
Prof. dr. ir. W.G.J. van der Meer	Delft University of Technology
Prof. dr. S.K. Hong	Korea University, South Korea
Prof. dr. T.O. Leiknes	KAUST/NTNU
Dr. ir. B. Blankert	Oasen, the Netherlands
Prof. dr. ir. L.C. Rietveld	Delft University of Technology (reserve member)

CRC Press/Balkema is an imprint of the Taylor & Francis Group, an informa business

Published by:
CRC Press/Balkema
PO Box 11320, 2301 EH Leiden, the Netherlands
e-mail: Pub.NL@taylorandfrancis.com
www.crcpress.com - www.taylorandfrancis.com

ISBN 978-1-138-02686-5 (Taylor & Francis Group)

Acknowledgements

My sincere gratitude goes to my supervisor, Prof. Maria D. Kennedy, for giving me the opportunity to pursue my academic career and for her patience, support and encouragement throughout the research period and particularly during the long search for funding. I am forever grateful to Prof. Jan C. Schippers for his guidance, encouragement, support and critical discussions without which the completion of this thesis would have been difficult if not impossible.

Thanks are due to the Ministry of Economic Affairs of the Netherlands for co-funding this research project. This thesis was part of the project "Zero Chemical UF/RO for Seawater Desalination" in collaboration with 5 industrial and academic partners; Pentair X-flow, Vitens, Evides, Twente University and RWTH Aachen. I would like to thank Pentair for supporting this study and for providing materials and equipment. The water supply company Evides was very gracious in providing a test ground at the demonstration desalination plant in Zeeland.

The laboratory staff at UNESCO-IHE lent their invaluable support during this period with analyses, materials, equipment, calibrations, re-calibrations, and a never-ending list of activities. For this I am grateful to Fred Kruis, Peter Heerings, Lyzette Robbemont, Ferdi Battes, Frank Wiegman and Don van Galen.

My thanks go to Mr. Jan-Herman Koster for providing me with the opportunity to engage in a capacity building project in the water and sanitation sector in Iran for two years while waiting for the PhD funding to come through. I am grateful to Klaas Schwartz for generously involving me in his projects, against all odds.

I would like to thank Loreen Villacorte for being a great colleague and peer and for his ever-ready support throughout this period. The insightful discussions with Bastiaan Blankert, Rinnert Schurer, Ferry Horvath, Frederik Spenkelink and Paul Buijs were extremely helpful in shaping my understanding of the topic at large and connecting my work to practice.

The contribution of a number of colleagues was crucial for the completion of this work. I would like to thank in alphabetic order, Abubakar Hassan, Bayardo Gonzalez, Clement Trellu, Helga

Calix, Madiany Hernandez, Mari Gonzalez, Marion Joulie, Mohaned Sousi, Nirajan Dhakal, Peter Mawioo, Rohan Jain, Tom Spanjer, Yasmina Bennani and Yuli Ekowati.
Thanks are due to Dr. Stefan Huber, Dr. Andreas Balz and Dr. Michael Abert of DOC-Labor for their ever ready presence and support with LC-OCD analyses. I would like to express my gratitude to Kaisa Karisalmi, Andreja Peternelj and Mehrdad Hessampour for the fruitful discussions.

The bonds that have formed with a number of friends over the past so many years are no less than familial and have certainly proven to me that in loving and caring, biology is over-rated! Maria Pascual thank you for being such an amazing friend, for your endless kindness, compassion, positivity and joy. Maria Rusca, one day I will build a statue of you and put it where all can see. For now, I keep you very closely in my heart! Stefania, your friendship is beyond gold. And you my dear are simply irreplaceable! Denys, no matter how far the physical distance, I will always be there for you and my unofficial god-child with all my heart and soul! Mona, my admiration for you is endless. Thank you for being there for all these years. Nora, your timing was impeccable, your presence has made all the difference, and I daresay finishing this work would have been very difficult without you.

To the people who daily, weekly, monthly, or from time to time enriched my life with their existence; Saul Buitrago, Juan Pablo Aguilar Lopez, Jorge Almaraz, Emmanuel Mulo Ogwal, Sergiu Chelcea, Ina Kruger, Andreas Plischke, Patricia Trambauer, Guy Beaujot, Joe Neesan, Davide Merli, Giuliana Ferrero, Joana Cassidy, Paloma Ve, Lakshmi Charli, Angelica Rada, Ghazaleh Jasbi, Mirena Olaizola, Soudabeh Rajaei. Thank you!

The support, love and encouragement of my parents has been more than I could have asked for. Salma and Sara, the most amazing sisters on the planet; Reza, a true role model and amazing brother; Mo, an incredible artist and scholar; it would have been impossible to go through these last couple of years if it was not for your encouragement and inspiring presence.

I am forever grateful to Mireille Lambert, Karen Lind, Catayoun Azmayesh, Leila Azmayesh, Bijan Azmayesh and Damien Lallemand for being an endless source of love and an immense support since I started this journey. To my uncles Ali Mohammad Azmayesh and Ali Asghar Azmayesh, thank you for believing in me and for making sure that I would have all I needed to achieve this goal. It's done! We did it!

To my biological and non-biological family

Summary

Seawater desalination is a globally expanding coastal industry with an installed capacity of 80 million m^3/day as of 2013. Reverse osmosis (RO) has become the dominant technology for seawater desalination with more than two thirds of the global installed desalination capacity. The major challenge for cost-effective application of seawater RO (SWRO) systems is membrane fouling. To mitigate fouling and reduce associated operational problems, pretreatment by granular media filtration (GMF) or micro- and ultrafiltration (MF/UF) is commonly required.

Operation of SWRO pretreatment has proven to be challenging during algal bloom periods where relatively high concentrations of algal cells and algal organic matter (AOM) are present in seawater. Experience from a severe red tide bloom in the Middle East in 2008-2009 showed that GMF in combination with coagulation cannot handle severe algal bloom events. During this period, the failure of GMF to produce acceptable RO feed water quality (silt density index, SDI < 5) caused the shutdown of several desalination plants in the region. This event highlighted the importance of reliable pretreatment systems for SWRO operation, and focused the attention of the desalination industry on MF/UF technology.

MF/UF systems are generally more reliable than GMF in producing stable, high quality RO feed water in terms of turbidity and SDI. Moreover, MF/UF product water quality is not affected by variations in raw water quality. Experience with large-scale UF operation in SWRO pretreatment during severe algal bloom events is limited and data is scarce. However, a well documented case of UF/RO operation on North Sea water in the Netherlands showed that during severe algal bloom periods, coagulation was required to stabilize UF hydraulic performance. In general, MF/UF membranes do not rely on coagulation to reduce turbidity and SDI. However, coagulation may enhance AOM removal in MF/UF systems and reduce particulate/organic and biofouling potential of UF permeate. From an operational point of view, it is desirable to completely eliminate coagulation from the process scheme, to reduce costs and complexities associated with chemicals, waste treatment, handling and discharge.

The goal of this study was to evaluate the feasibility of UF as pretreatment to SWRO during algal bloom periods and to investigate the role of coagulation in improving UF operation. Ultimately the study aimed at providing insight into options for minimizing and ideally eliminating coagulation from UF pretreatment to SWRO.

Algal blooms adversely affect UF operation by causing higher pressure development during filtration; lower permeability recovery after backwashing; and high concentration of algal biopolymers in UF permeate. The latter results in higher particulate/organic and biofouling potential of SWRO feed water.

The first step of the study was to understand particle properties that affect fouling in MF/UF systems. Theoretical calculations indicated that spherical particles as small as a few nanometres - forming cake/gel layers with porosity ranging from 0.4 to 0.99 - do not contribute significantly to pressure increase in MF/UF systems operated at constant flux, indicating that the creation of large aggregates by e.g., extended flocculation, is not required in these systems.

Further investigations were made to study the effect of process conditions on inline coagulation with ferric chloride prior to MF/UF. Experimental results showed that extended flocculation was not required for inline coagulation prior to MF/UF systems treating surface water and proper selection of dose and pH was sufficient to optimize MF/UF operation in terms of fouling potential and permeate quality. Calculations indicated that high G-values and short residence times encountered prior to and within MF/UF elements in practice, seem to be sufficient to maintain low fouling potential and control non-backwashable fouling.

The effect of coagulation on hydraulic performance and permeate quality of UF membranes fed with AOM solutions in synthetic seawater was investigated. AOM biopolymers had high fouling potential as measured by the Modified Fouling Index (MFI) and were very compressible. Filtration at higher flux exacerbated both fouling potential and compressibility of AOM. Coagulation substantially reduced fouling potential, compressibility and flux dependency of AOM, resulting in substantially lower pressure development in filtration tests at constant flux. Inline coagulation/UF was more effective than conventional coagulation followed by filtration (0.45 μm) in terms of biopolymer removal at low coagulant dose (~ 0.5 mg Fe(III)/L).

The applicability of coating UF membranes with a removable layer of particles at the start of each filtration cycle for treating algal bloom-impacted seawater was investigated. Iron hydroxide particles were applied as coating material at the start of each filtration cycle at different equivalent dose. Without coating, AOM filtration was characterized by poor backwashability. Pre-coating was effective in controlling non-backwashable fouling using ferric hydroxide prepared by simple precipitation and low intensity grinding. However, relatively high equivalent dose (~ 3 - 6 mg Fe(III)/L) was applied. Pre-coating with ferric hydroxide particles smaller than 1 μm - prepared by precipitation and high intensity grinding - resulted in low equivalent dose of 0.3 - 0.5 mg Fe(III)/L required for stable operation of the UF membranes. Further reducing particle size of the coating material is expected to be more effective in lowering the required equivalent dose. However, preparation of such particles requires further research efforts.

Coagulation of AOM was studied for conventional coagulation (coagulation/flocculation and sedimentation) followed by filtration (0.45 μm), to identify AOM removal rates in seawater. Coagulation followed by sedimentation required coagulant dose of up to 20 mg Fe(III)/L to remove AOM biopolymers by up to 70%. Filtration through 0.45 μm had a significant impact on AOM removal, even at coagulant dose < 1 mg Fe(III)/L. This indicated that coagulated AOM aggregates have better filterability than settlability characteristics which may be attributed to the low density of these aggregates and could have considerable implications for the choice of clarification process in conventional pretreatment systems.

The applicability of low molecular weight cut-off UF (10 kDa) membranes as a coagulant-free alternative to SWRO pretreatment was investigated. 10 kDa membranes were capable of completely removing AOM biopolymers from SWRO feed water without coagulation. UF membranes with nominal molecular weight cut-off of 150 kDa reduced biopolymer concentration to approximately 200 μg C/L (~ 60% removal). In terms of hydraulic operation, 10 kDa membranes showed lower permeability recovery after backwash than 150 kDa membranes. Physical characterization of the two membranes revealed much lower surface porosity of 10 kDa compared to 150 kDa membranes.

In general terms, this study demonstrated that during algal blooms, UF membranes with nominal molecular weight cut-off of 150 kDa operated in inside-out mode, are more capable of reducing particulate/organic fouling potential of SWRO feed water at low coagulant dose than conventional coagulation. The application of UF membranes with low molecular weight cut-off can further enhance RO feed water quality in terms of particulate/organic fouling potential during algal blooms, without the need for coagulation. However, a small amount of coagulant may be required to control hydraulic operation of the UF membranes during these periods. Further improvements in material properties of these membranes should be directed at increasing the surface porosity of the membranes to enhance permeability recovery and ensure stable hydraulic operation.

Samenvatting

Zeewater ontzilting is een kustindustrie die wereldwijd uitbreidt met een geïnstalleerde capaciteit van 80 miljoen m^3/dag. Verantwoordelijk voor meer dan tweederde van de wereldwijd geïnstalleerde ontziltingcapaciteit, is omgekeerde osmose (RO) de dominante technologie geworden voor zeewater ontzilting. De grootste uitdaging voor een kost-effectieve toepassing van zeewater RO (SWRO) is membraan fouling. Om membraan fouling en de geassocieerde operationele problemen tegen te gaan, is doorgaans voorbehandeling vereist met granulair medium filtratie (GMF) of micro- en ultrafiltratie (MF/UF).

Tijdens periodes van algenbloei, wanneer er relatief hoge concentratie aan cellen en organische materie van algen (AOM) in het voedingwater aanwezig zijn, is de werking van voorbehandeling in SWRO een uitdaging gebleken. Ervaring met een ernstige "red tide" algenbloei in het Midden-Oosten in 2008-2009 heeft aangetoond dat GMF in combinatie met coagulatie ernstige algenbloei niet aankan. Tijdens deze periode, heeft het falen van GMF om RO voedingwater van aanvaardbare kwaliteit te produceren (silt density index, SDI > 5) verschillende ontziltinginstallaties in de regio stilgelegd. Deze gebeurtenis heeft het belang van betrouwbare voorbehandelingsystemen voor de werking van SWRO aangetoond en heeft de aandacht van de ontziltingindustrie op MF/UF technologie gevestigd.

MF/UF systemen zijn algemeen betrouwbaarder dan GMF om RO voedingwater van stabiele en hoge kwaliteit te produceren in termen van troebelheid en SDI. Daarbovenop wordt de geproduceerde waterkwaliteit van MF/UF niet aangetast door variaties in de kwaliteit van het ruwe water. Algemeen gesproken, zijn MF/UF membranen niet afhankelijk van coagulatie om troebelheid en SDI te reduceren. Met de werking van UF in SWRO voorbehandeling op grote schaal is er slechts beperkte ervaring en er zijn weinig data beschikbaar. Een goed gedocumenteerd geval van UF/RO werking op Noordzee water in Nederland toonde echter aan dat coagulatie vereist is gedurende periodes van ernstige algenbloei om de hydraulische prestatie van UF te stabiliseren. Bovendien kan coagulatie AOM verwijdering in MF/UF systemen verhogen en het fouling potentieel onder de vorm van partikels/organische- of biofouling van het UF permeaat reduceren. Vanuit een operationeel oogpunt is het wenselijk om coagulatie volledig uit het processchema te halen, om zo kosten en complicaties geassocieerd met chemische stoffen en behandeling, omgang en lozing van afval te reduceren.

Het doel van deze studie was om de haalbaarheid van UF als voorbehandeling van SWRO tijdens periodes van algenbloei te evalueren op het vlak van hydraulische prestatie en kwaliteit van het productwater, door enerzijds de rol van coagulatie in de hydraulische prestatie van UF en anderzijds de permeaatkwaliteit te bestuderen. Tenslotte, doelde de studie op het aanbrengen van meer inzicht in mogelijkheden om coagulatie in op UF gebaseerde voorbehandeling van SWRO te verminderen en idealiter te verwijderen.

Algenbloei beïnvloedt UF werking negatief door het veroorzaken van; hogere drukontwikkeling tijdens filtratie, lagere permeabiliteitherstel na terugspoelen, en hogere concentratie aan algen biopolymeren in het UF permeaat. Dit laatste resulteert in een hoger partikel/organische en biofouling potentieel van het SWRO voedingwater.

Een eerste stap in deze studie was om te begrijpen welke deeltjeseigenschappen invloed hebben op membraan fouling in MF/UF systemen. Theoretische berekeningen toonden aan dat sferische deeltjes met afmetingen van een paar nanometer – die cake/gel lagen vormen met een porositeit van 0.4 tot 0.99 – niet significant bijdragen tot druktoename in MF/UF systemen die met constante flux geopereerd worden, wat aangeeft dat het niet nodig is om in deze systemen grote aggregaten te creëren door bv. uitgebreide flocculatie.

Verder onderzoek werd gedaan om het effect te bestuderen van procesvoorwaarden op inline coagulatie met ijzerchloride, stroomopwaarts van MF/UF. Experimentele resultaten tonen aan dat uitgebreide flocculatie niet vereist was voor inline coagulatie stroomopwaarts van MF/UF systemen die oppervlaktewater behandelen en dat een geschikte selectie van dosis en pH voldoende was om MF/UF werking te optimaliseren op gebied van permeaatkwaliteit en het potentieel voor membraan fouling. Berekeningen tonen aan dat hoge G-waarden en korte residentietijden stroomopwaarts van en in MF/UF elementen in werking, voldoende blijken om een laag fouling potentieel te behouden en niet-terugspoelbare fouling te controleren.

Deze studie onderzocht het effect van coagulatie op hydraulische prestatie en permeaatkwaliteit van UF membranen gevoed met AOM oplossingen in synthetisch zeewater. AOM biopolymeren hadden een hoog fouling potentieel gemeten met de Modified Fouling Index (MFI) en waren erg samendrukbaar. Filtratie met hogere flux verscherpte zowel fouling potentieel als samendrukbaarheid van AOM. Coagulatie reduceerde de fouling potentieel, samendrukbaarheid en flux-afhankelijkheid van AOM substantieel, resulterend in substantieel lagere drukontwikkeling in filtratietesten bij constante flux. Inline coagulatie/UF was effectiever dan conventionele coagulatie gevolgd door filtratie (0.45 μm) op gebied van biopolymeer verwijdering bij lage dosis coagulant (~ 0.5 mg Fe(III)/L).

De toepasbaarheid van het coaten van UF membranen met een verwijderbare laag partikels bij de start van elke filtratiecyclus voor de behandeling van door algenbloei beïnvloed zeewater werd onderzocht. Ijzerhydroxide deeltjes werden toegediend als coating materiaal bij de start van elke filtratiecyclus bij verschillende equivalent doses. Zonder coating werd AOM gekenmerkt door een slechte terugspoelbaarheid. Pre-coating was effectief om niet-terugspoelbare fouling te controleren, met gebruik van ijzerhydroxide, die bereid werd door simpele neerslag en mixen op lage intensiteit. Er werden echter relatief hoge equivalent doses (~ 3 - 6 mg Fe(III)/L) toegediend. Pre-coating met ijzerhydroxide deeltjes kleiner dan 1 μm – bereid door neerslag en mixen op hoge intensiteit – resulteerde in lage equivalent doses van 0.3 - 0.5 mg Fe(III)/L nodig voor een stabiele werking van de UF membranen. Het is verwacht dat een verdere reductie van deeltjesafmetingen van het coating materiaal effectiever is in het verlagen van de vereiste equivalent dosis. De bereiding van zulke deeltjes vereist verder onderzoek.

Coagulatie van AOM werd bestudeerd voor conventionele coagulatie (coagulatie/flocculatie en sedimentatie) gevolgd door filtratie (0.45 μm), om de verwijderinggraad van AOM in zeewater te identificeren. Coagulatie gevolgd door sedimentatie vereist een coagulant dosis tot 20 mg Fe(III)/L om AOM biopolymeren tot 70% te verwijderen. Filtratie door 0.45 μm had een significante invloed op AOM verwijdering, zelfs bij coagulant dosis < 1 mg Fe(III)/L. Dit toont aan dat gecoaguleerde AOM aggregaten betere filterbaarheid dan bezinkbaarheid karakteristieken hebben, wat kan toegewezen worden aan de lage densiteit van deze aggregaten. Dit kan aanzienlijke implicaties hebben voor de keuze van het klaringsproces in conventionele voorbehandelingsystemen.

De toepassing van UF membranen van laag poriegrootte of molecular weight cut-off (10 kDa) als een coagulantvrij alternatief voor SWRO voorbehandeling werd bestudeerd. 10 kDa membranen waren in staat om AOM biopolymeren volledig uit SWRO voedingwater te verwijderen zonder coagulatie. UF membranen met een nominaal molecular weight cut-off van 150 kDa reduceerden biopolymeer concentraties tot ongeveer 200 μg C/L (~ 60% verwijdering). Op gebied van hydraulische prestaties, toonden 10 kDa membranen een lagere permeabiliteit na terugspoel dan de 150 kDa membranen. Fysieke karakterisering van de twee membranen toonde een veel lagere oppervlakteporositeit aan voor de 10 kDa membranen in vergelijking met de 150 kDa membranen.

Globaal gesproken, heeft deze studie aangetoond dat UF membranen met een nominale molecular weight cut-off van 150 kDa geopereerd in inside-out modus beter in staat zijn om partikel/organische fouling potentieel van SWRO voedingwater te reduceren bij lagere coagulant dosis dan conventionele coagulatie. De toepassing van UF membranen met lage molecular weight cut-off kan de RO voedingwater kwaliteit verder verbeteren op gebied van partikel/organische fouling potentieel tijdens algenbloei, zonder noodzaak tot coagulatie. Een kleine hoeveelheid coagulant kan echter vereist zijn om de hydraulische werking van UF membranen tijdens deze periodes te controleren. Verdere verbeteringen in materiaaleigenschappen van deze membranen zouden gericht moeten worden op het verhogen van oppervlakteporositeit van deze membranen om het herstel van permeabiliteit te verhogen en om een stabiele hydraulische werking te garanderen.

Contents

1

GENERAL INTRODUCTION

1.1 Background

Increasing demand for safe water supply, due to population growth and economic development, has led to depletion and deterioration of existing fresh water resources and water stress in many parts of the world. It is estimated that by 2050, 5 billion of the projected world population of 9.7 billion people will live in areas of high water stress, while another 1 billion will be living in areas of water scarcity (Schlosser et al., 2014). To satisfy demand, many communities are implementing management schemes to increase consumption efficiency; transporting water over large distances; or resorting to alternative sources for water supply such as water reuse and desalination of brackish water and seawater. With half of the world population living within 60 km of the sea (UNEP, 2014) and many countries in the Middle East, North Africa and Asia already facing serious water scarcity (WWAP, 2012), seawater desalination is increasingly applied to satisfy the demand for safe water supply of adequate quantity.

Common technologies for desalination can be classified into distillation techniques such as multistage flash (MSF), multi effect distillation (MED) and vapour compression distillation; and membrane-based techniques such as nanofiltration (NF), reverse osmosis (RO) and electro dialysis (ED). Distillation techniques are mainly applied for desalination of seawater, particularly in the Middle East, while membrane-based techniques were initially successfully introduced for fresh and brackish water treatment. However, in the last two decades RO has found its application for seawater desalination as well, and has outperformed distillation techniques.

With more than two-thirds of the global installed desalination capacity (DesalData, 2014), reverse osmosis has become the dominant technology for production of fresh water from saline water sources. This is mainly due to reduced energy requirements, process simplicity in design and compactness (Brehant et al., 2002) compared to other desalination technologies. In the last 20 years of development, RO performance has improved markedly and prices have decreased significantly owing to market expansion and larger projects (Pearce, 2007a). The current water production cost of RO desalination is generally cheaper than thermal desalination processes and it is expected that by 2015, average production cost of RO desalinated water will be approximately 0.5 USD/m^3, making large scale application of SWRO economically attractive (GWI, 2007).

The main challenge facing RO systems operation is membrane fouling which affects process efficiency both in terms of quality and quantity of produced water (Matin et al., 2011). In general suspended and colloidal particles, organic matter, dissolved nutrients and sparingly soluble salts, can cause fouling and scaling in RO membranes. Fouling may result in frequent chemical cleaning, long downtime, shorter membrane lifetime, higher energy consumption and lower product water quality (Wilf and Schierach 2001; Hoof et al., 2001). In seawater RO (SWRO) systems algal blooms can enhance various forms of fouling due to increased total suspended solids and organic content resulting from algal biomass in the raw water (Caron et al., 2010).

Pretreatment is commonly required to mitigate fouling in RO systems, ensure adequate membrane productivity and extend membrane lifetime.

1.2 Pretreatment for seawater reverse osmosis

Most RO system failures are the result of poorly designed or operated pretreatment systems (Gallego and Darton, 2007). Pretreatment normally involves a form of filtration and other physical-chemical processes whose primary purpose is to remove suspended solids (particles, silt, algae, organics, etc.), oil and grease from the source water (WHO, 2007). Pretreatment is generally categorized as primary pretreatment, i.e., coagulation and clarification processes such as sedimentation and dissolved air flotation (DAF); and secondary pretreatment, i.e., filtration processes. Secondary pretreatment is classified as conventional when granular media filters (GMF) are applied and advanced when micro- and ultrafiltration (MF/UF) membranes are used. Large differences in water composition at different sources and locations make design, optimization and operation of pretreatment systems quite a challenge.

Cumulative installed capacity per pretreatment technology (GMF, UF and DAF) for the top 50 SWRO plants in terms of capacity, installed between 2001 and 2013, is presented in Figure 1-1. It must be noted that DAF is a primary pretreatment process that is normally applied in combination with either GMF and/or UF systems. The last decade has seen a rise in both UF and DAF technology in SWRO pretreatment.

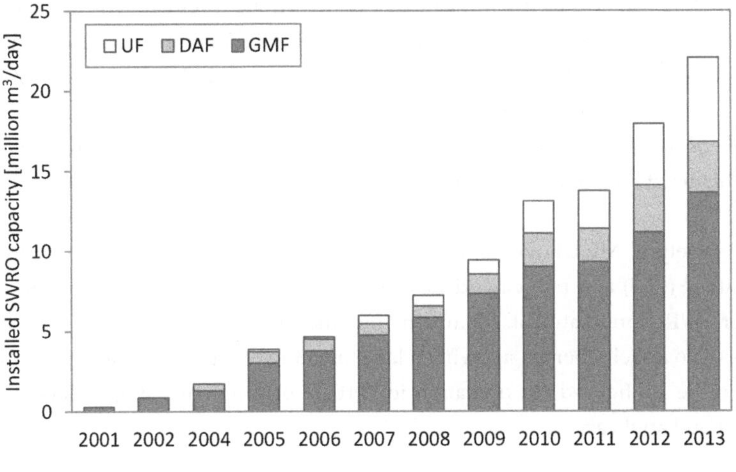

Figure 1-1 SWRO plant capacity in terms of pretreatment technology, GMF, UF and DAF. Primary data from DesalData (2013)

1.2.1 Conventional pretreatment systems

In conventional systems, a variety of configurations of primary and secondary pretreatment have been applied in SWRO desalination plants, from no pretreatment to extensive primary and secondary pretreatment consisting of coagulation, flocculation, sedimentation (e.g., Trinidad) or DAF (e.g., Tuas, Singapore), and media filtration in dual-stage (Busch et al., 2010).

Conventional pretreatment of seawater having low fouling potential is typically based on granular media filtration systems, such as gravity or pressurized dual media filters (DMF) - typically sand and anthracite - in single or dual stage. Media filtration - treating water with higher fouling potential - relies on coagulation to enhance the removal of fine particles and dissolved organics (Pearce, 2009). Coagulation is applied in inline mode (direct filtration) having very little or no flocculation time. More severely contaminated sources need coagulation and flocculation followed by a clarification step such as sedimentation or DAF prior to media filtration. When raw water quality is poor, high coagulant doses are required, resulting in a high concentration of coagulated particles. Accumulation of these particles at the surface of granular media filters can cause high head loss and therefore, removal of particles through clarification (by sedimentation or flotation) is necessary.

1.2.2 Advanced pretreatment systems

For over a decade MF/UF membranes have been tested and applied at pilot and commercial scale as pretreatment for SWRO (Wilf and Schierach, 2001; Glueckstern et al., 2002; Brehant et al., 2002; Wolf et al., 2005; Halpern et al., 2005; Gille and Czolkoss, 2005). MF/UF membranes offer several advantages over conventional pretreatment systems; lower footprint, constant high permeate quality in terms of silt density index (SDI), higher retention of large molecular weight organics, lower overall chemical consumption, etc. (Wilf and Schierach, 2001; Pearce, 2007a). Successful piloting has led to the implementation of MF/UF pretreatment in several large (> 100,000 m^3/day) SWRO plants (Busch et al., 2010). As of 2013, the total installed SWRO capacity with MF/UF as pretreatment is about 5 Mm^3/day (Figure 1-2). MF/UF as pretreatment to SWRO is applied in many configurations, from facilities providing MF/UF preceded by various combinations of primary and secondary conventional treatment processes (Basha et al., 2011) to facilities using direct membrane filtration.

There are a variety of MF/UF membrane materials and configurations commercially available. Polyethersulfone (PES) and polyvinylidene difluoride (PVDF) are the two dominant membrane materials for SWRO pretreatment. Membrane modules are made by packing several thousand hollow fibres into a shell, thereby allowing a large membrane area per square meter of footprint. Modules may be configured for operation in vertical or horizontal mode, depending on the manufacturer and application.

In broad terms, the products can be separated into two groups based on different flow modes; inside-out and outside-in. The pressure differential driving force is another distinguishing characteristic; pressurizing the membranes on the feed side vs. applying a vacuum on the permeate side of the membrane (submerged systems). PES membranes are commonly operated in pressure-driven inside-out filtration mode, whereas PVDF is either pressure-driven or submerged with outside-in filtration.

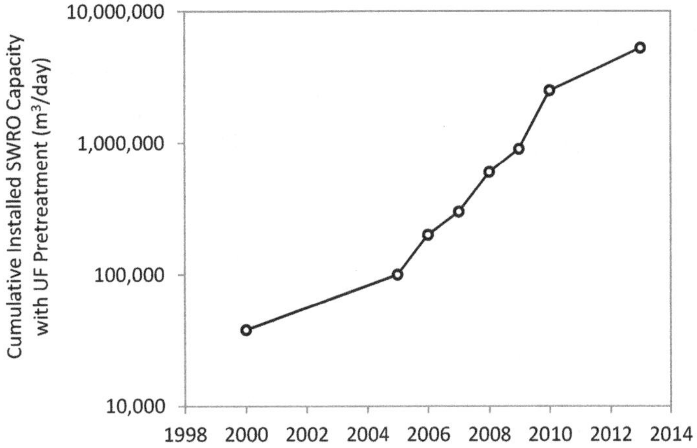

Figure 1-2 Cumulative installed SWRO capacity with UF as pretreatment (Figure adapted from Busch et al., 2010; primary data for 2011-2013 from DesalData, 2013)

In inside-out filtration, feed water enters the membranes from the lumen. Hydraulic cleaning is performed by reversing the flow, whereby product water flows through the membranes from the outside of the hollow fibres physically lifting the fouling material from the membrane surface and flushing it out of the lumen. In outside-in configuration, feed water enters from the outside of the fibre lumen and product water flows through the inside of the fibre. Hydraulic cleaning is performed by flow reversal, whereby product water flows from the inside of the fibres through to the outside, cleaning fouling material off from the outer part of the fibres. Outside-in configurations usually incorporate an air scour during backwashing to facilitate hydraulic cleaning.

Each material and configuration has its advantages. For a given fibre, the area based on the outside diameter is typically larger than the area based on the inside diameter. Inside-out fibres tend to have higher permeability, due to the selection of PES rather than PVDF (Pearce, 2007). An overview of 10 international membrane suppliers for SWRO pretreatment is provided in alphabetic order in Table 1-1. In recent years and with the growth in market for SWRO pretreatment, many membrane manufacturers have adapted their products to specifically cater to the seawater market.

Hydranautics, inge and Pentair X-Flow exclusively manufacture inside-out PES membranes. Hyflux supplies both PES and PVDF membrane fibres. The other 6 manufacturers presented in Table 1-1 supply PVDF membranes for seawater RO pretreatment. PES has seen a substantial growth in installed capacity due to several large projects, e.g., Ashdod, Magtaa, Shuwaikh, etc.

Table 1-1 Key products of 10 international MF/UF membrane manufacturers for SWRO pretreatment

Manufacturer	Product	Type	Material	Configuration	
Dow	IntegraPAC™ IntegraFlo™	UF	PVDF	Outside-in	Pressurized
Hydranautics	Hydracap®	UF	PES	Inside-out	Pressurized
Hyflux	Krital® 600E Kristal® 2000E	UF	PES PVDF	Outside-in	Pressurized
inge	Multibore®	UF	PES	Inside-out	Pressurized
Koch	PURON HF	UF	PVDF	Outside-in	Submerged
Siemens	Memcor®	UF	PVDF	Outside-in	Pressurized
Pall-Asahi	Microza™ UNA	MF	PVDF	Outside-in	Pressurized
Pentair X-Flow	Seaguard Seaflex	UF	PES	Inside-out	Pressurized
Toray	TORAYFIL®	UF	PVDF	Outside-in	Pressurized Submerged
GE-Zenon	Zeeweed® 1000 Zeeweed® 1500	UF	PVDF	Outside-in	Submerged Pressurized

UF membranes have wider application for seawater pretreatment than MF membranes mainly due to better removal of suspended organics, silt and pathogens from seawater, as UF membrane pores are significantly smaller than those of the MF membranes (Voutchkov, 2009). This study focuses on single bore, inside-out PES hollow fibre UF membranes for SWRO pretreatment.

1.3 Chemical consumption in SWRO pretreatment

To ensure that RO membranes and pretreatment systems operate smoothly and efficiently, chemicals are required. Chemicals increase the cost and complexity of the overall process. The chemicals required for RO operation do not depend on pretreatment type. These may include sodium bisulphite to neutralize chlorine (chlorine may be used intermittently to disinfect the intake system of SWRO plants and reduce marine growth); antiscalants and/or acid for scale inhibition; chemicals for cleaning in place (CIP) such as alkaline solutions to remove silt deposits and biofilms, acid to dissolve metal oxides, detergents or oxidants; and disinfection chemicals such as hydrogen peroxide or DBNPA (WHO 2007; Lattemann, 2010).

Chemicals used in SWRO pretreatment are similar for both conventional and advanced pretreatment systems and only differ in dose, with the exception of sodium hypochlorite and sodium hydroxide that are exclusively used in MF/UF systems for chemically enhanced backwashing (CEB) of the membranes.

Coagulation in conventional pretreatment systems is mainly applied to assure that product water quality of GMF meets the requirements of RO membrane manufacturers in terms of turbidity and SDI. The most commonly used coagulant in SWRO pretreatment is ferric chloride. Aluminium based coagulants are generally not used due to the high solubility of aluminium that could result in carryover to the RO membranes where it can concentrate causing precipitative scaling by aluminium hydroxide and aluminium silicate (Edzwald and Haarhoff, 2011). Coagulant dose in conventional pretreatment systems may range from 0.5 to 10 mg Fe/L, although in some cases doses as high as 20 mg Fe/L have been reported (Lattemann, 2010; Edzwald and Haarhoff, 2011).

Coagulant aids are high molecular weight anionic or nonionic polymers (e.g., polyacrylamides) that are added after coagulation, at dosages of 0.2-2 mg/L, to enhance floc growth and floc strength prior to sedimentation. They may also be used to enhance floc strength in flotation and as a filter aid for GMF to aid attachment of flocs to filter aids (Lattemann, 2010; Edzwald and Haarhoff, 2011).

MF/UF systems do not rely on coagulation to enhance permeate quality in terms of turbidity and silt density index (SDI). However, coagulant (usually ferric chloride) is dosed at concentrations of 0.1 - 3 mg/L (Busch et al., 2010) to control non-backwashable fouling. Coagulant dose is normally correlated to feed water quality (Lattemann, 2010). Coagulation can help reduce pressure development in subsequent filtration cycles by enhancing cake filtration mechanism. Coagulation can also reduce the extent of non-backwashable fouling in UF systems by reducing pore blocking and/or surface attachment of sticky particles (Schurer et al., 2013).

Cleaning of MF/UF membranes requires chemicals for CEB and cleaning in place (CIP). Cleaning frequency depends on raw water quality and manufacturer recommendation. Some manufacturers recommend CEB when a certain permeability level has been reached. Others consider filtration time as the trigger for chemical cleaning. As such, CEB may be required as frequently as 1-2 times per day, while CIP may be performed a few times per year (WHO, 2007). CEB is performed by backwashing and soaking the membranes with sodium hypochlorite at high pH to remove biological material followed by acid rinsing to dissolve precipitated calcium carbonate. CIP is commonly performed with specialty cleaning chemicals with proprietary formulae, EDTA, etc. Tailor-made CIP recipes have been applied by different plant operators and/or membrane manufacturers depending on fouling type and extent.

Potential impacts on the environment associated with pretreatment chemicals (e.g., coagulants, coagulant aids) and their disposal (WHO, 2007) increase process complexity. Chemical discharge to the marine environment may have adverse effects on water and sediment quality, and can impair marine life and functioning of coastal ecosystems (Lattemann and Höpner, 2008). Coagulants (e.g., ferric chloride) and coagulant aids (high molecular weight organics e.g., polyacrylamide) present in spent backwash water are typically discharged to the ocean without treatment (Lattemann, 2010). The coagulant itself has a low toxic potential. However, discharge may cause an intense coloration of the reject stream if ferric salts are used (red brines), which may increase turbidity and reduce light penetration, or could bury sessile benthic organisms in the discharge site (Latteman and Höpner, 2008). Coagulant aids and chlorine are highly toxic compounds and their toxicity has been investigated in several studies.

In areas with more stringent legislation on concentrate discharge, such as Europe, Australia and USA, coagulant-rich waste streams require extensive treatment and handling prior to discharge, which pose a significant cost component to the overall pretreatment process. In such cases backwash water containing chemicals is treated separately by e.g., gravity settling in lamella plate sedimentation tanks. Supernatant can be either disposed with RO concentrate or recycled at the head of the pretreatment. The coagulant-rich sludge retained in the sedimentation tank is often dewatered onsite and transported to sludge treatment facilities or landfills (WHO, 2007).

Given the complexities associated with treatment and disposal of coagulant-rich waste, operators give preference to systems operating with no coagulant or if not evitable, systems that require minimum amounts of such chemicals.

1.4 Algal blooms and seawater reverse osmosis operation

The severe red tide event that occurred in the Middle East in 2008-2009 and led to the shutdown of several desalination plants in the region highlighted the risk that algal blooms pose for desalination plants and drew specific attention to the phenomenon.

Algal blooms occur when phytoplankton proliferate exponentially due to availability of nutrients and sunlight, and continue to increase rapidly in cell numbers until nutrients are depleted. Algal blooms are seasonal phenomena and are commonly observed in spring - temperate climates - when light and temperature favour the growth of algal cells. Although algal blooms can be beneficial to the food chain in aquatic environments by recycling nutrients (Falkowski et al., 1998), some algal blooms are detrimental to the environment and humans due to their high biomass concentrations and through release of toxins (the high biomass concentrations of algae, during algal bloom conditions can negatively impact other species). Adverse effects include water discolouration, light and oxygen depletion, and release of toxins that can alter cellular processes and cause mortalities in aquatic organisms and mammals (Sellner et al., 2003). Toxin-producing algal blooms are commonly referred to as harmful algal blooms, of which red tide events are an example. The rise in frequency and geographic distribution of algal blooms in the past decade is mainly attributed to improved monitoring and advanced recording techniques of the occurrence of such phenomena (Anderson et al., 2012).

The challenges of treating algal bloom-impacted water in seawater RO desalination plants are two-fold. The first challenge is to ensure that toxins are removed by the RO membranes. Laboratory studies show that RO membranes can remove > 99% of algal-derived toxins (Laycock et al., 2012). The other challenge is the effect of algal-derived organic matter on the operation of pretreatment systems and RO membranes.

Algal blooms produce different forms and varying concentrations of algal organic matter (AOM). AOM comprises intracellular organic matter (IOM) formed due to autolysis consisting of proteins, nucleic acids, lipids and small molecules as well as extracellular organic matter (EOM) formed via metabolic excretion and composed mainly of polysaccharides (Fogg, 1983; Myklestad, 1995). A significant fraction of these exo-polysaccharides that are highly surface-active and sticky and play an important role in the aggregation dynamics of algae during bloom events are known as transparent exopolymer particles (TEP) (Alldredge et al., 1993; Myklestad, 1995; Mopper et al., 1995).

Berman and Holenberg (2005) were the first to report that TEP can potentially initiate and exacerbate biofouling in RO systems. Villacorte (2014) investigated the effect of AOM including TEP in relation to UF fouling propensity. A number of other studies have investigated the effect of algal blooms on MF/UF system operation at laboratory scale (Kim and Yoon, 2005; Ladner et al., 2010; Qu et al., 2012) and SWRO systems at pilot scale (Petry et al., 2007; Kim et al., 2007).

From literature, two documented instances of SWRO operation during severe algal bloom events were encountered. These events are discussed in more detail below. A summary of the key characteristics of the blooms, and pretreatment operation is given in Table 1-2.

1.4.1 Middle East

A severe red tide event in 2008 - 2009 in the Gulf of Oman and the Persian Gulf resulted in massive fish kill, huge economic loss and the shutdown of several desalination plants in the region. The bloom was caused by the dinoflagellate *Cochlodinium polykrikoides* with reported algal cell concentrations of up to 27,000 cells/mL (Richlen et al., 2010). Operation of SWRO plants during the red tide outbreak was marked by severe clogging of granular media filters resulting in short run lengths and reduced production capacity, which together with concerns about the quality of RO feed water (SDI > 5) led to the shutdown of several SWRO plants in the region. At the Fujairah plant in UAE, rapid clogging of the filters (lower productivity) together with lower product water quality forced the plant shutdown. Other plants did not disclose operational details. At least one distillation plant had to shut down due to taste and odour problems (J.C. Schippers, pers. comm., 2012).

1.4.2 Netherlands

Literature on operational experience with UF membranes as pretreatment to SWRO during periods of severe algal bloom is limited to one demonstration plant located at the Oosterschelde bay in the Netherlands operated from 2009 - 2013 (Schurer et al., 2012; 2013). The main species were identified as the haptophyte *Phaeocystis*, and the diatoms *Thalassiosira* and *Chaetoceros* with chlorophyll-a concentration of 60 μg/L in 2009. In 2009, algal bloom occurrence in the North Sea caused very high non-backwashable fouling of the UF membranes. The extent of fouling was such that in extreme conditions CEBs as frequent as once in 4 to < 12 hours were required (Schurer et al., 2012). Reducing the filtration flux was able to suppress UF fouling to some extent. However, even under these conditions sufficiently long CEB intervals could still not be re-established. Moreover, lower flux implied lower UF production capacity. Therefore, coagulation was required to control high fouling rates during spring algal bloom conditions.

Table 1-2 Overview of two SWRO desalination plants operation during severe algal bloom events, characteristics of the blooms, and pretreatment performance

	Middle East (2008/2009)	Netherlands (2009)
Plant capacity	170,000 m^3/d	360 m^3/d
Location	UAE, Gulf of Oman	Oosterschelde, North Sea
Pretreatment technology	DMF	UF
Coagulation	Continuous	Only during algal bloom
Bloom type	Toxic (red tide)	Non-toxic
Causative species	*Cochlodinium polykrikoides* (dinoflagellate)¤	*Phaeocystis* (haptophyte) *Thalssiosira* (diatom)§
Cell count	27,000 cells/mL¤	not known
Chl-a	not known	60 µg/L§
SDI of RO feed water	> 5‡	< 2§
Cleaning interval	Every 2 h (compared to every 24 hours outside bloom conditions)‡	Every 4-12 h in the peak of the bloom without coagulation (compared to every 2-7 d outside bloom conditions)§
Shutdown	Due to poor RO feed water quality‡,¤	coagulation was able to stabilize UF operation

¤ Richlen et al. (2010); ‡ Schippers, J.C. (2012); § Schurer et al (2013)

1.5 Need for research

MF/UF systems are rapidly gaining ground as pretreatment to SWRO. The reason behind this growth is that MF/UF pretreatment ensures low particulate fouling potential of feed water for RO membranes; reflected in SDI levels below 2-3 even for varying raw water quality. SDI is still the main parameter used in practice to judge pretreatment efficiency. Media filtration (including coagulation) has limited capacity in handling poor water quality and variations in quality. This is also true for extended pretreatment schemes including coagulation, flocculation and sedimentation prior to media filtration.

One of the most challenging operational regimes for SWRO pretreatment is during periods of algal bloom where relatively high concentrations of algal cells and algal organic matter (AOM) are present in seawater. Experience shows that conventional pretreatment systems (i.e., GMF in combination with coagulation) cannot handle severe algal bloom events, highlighting the need for more robust pretreatment technology. Some researchers (Edzwald, 2010) and practitioners (Le Gallou, 2011) have suggested the use of DAF prior to media filtration to overcome operational hurdles during periods of severe algal blooms. However, in some cases, DAF followed by single stage media filters may not be adequate and dual stage media filtration may be required (Petry et al., 2007). Moreover, chemical consumption in these systems is relatively high as coagulant is often dosed both upstream of DAF and prior to media filters. UF has been proposed as a solution for challenging operation during algal bloom periods.

Information and literature on large scale UF operation as SWRO pretreatment during severe algal blooms is scarce. Operational experience with a seawater UF/RO demonstration plant in the Netherlands showed rapid permeability decline and high fouling rates of the UF systems during algal bloom periods. Coagulation was able to stabilize UF operation. Coagulation entails process complexities and costs associated with chemicals, dosing equipment, storage capacity for chemicals, automation, etc. Moreover, if backwash water containing coagulant needs to be treated before discharge, higher investment and operational costs may be incurred for treatment, sludge handling and disposal. Given the complexities and costs associated with coagulation in UF/RO systems for seawater desalination, complete elimination of coagulation from these systems is highly desired.

To reduce or ideally eliminate coagulation from the process scheme of UF/RO seawater desalination plants it is imperative to understand factors affecting UF operation in SWRO pretreatment and the role of coagulation in enhancing UF process performance, and to find alternative solutions to minimize and ideally eliminate the need for coagulant in UF pretreatment to SWRO. This thesis proposes the following approach,

- Elucidate the effect of coagulation conditions on UF operation
- Investigate alternative modes of coagulant application (e.g., coating of UF membranes) that require minimal or no dose of metal-based coagulant to stabilize UF hydraulic performance
- Identify alternative membrane properties (e.g., pore size, surface porosity, etc.) that require less or no coagulant to remove AOM from RO feed water

The overall aim of this research is to provide insight into technical options for very low coagulant consumption in UF operation on seawater for the production of high quality feed for RO membranes.

1.6 Research objectives

The specific objectives of the proposed research are,

- To elucidate the effect of inline coagulation conditions (e.g., dose, pH, G, Gt) and filtration flux on AOM fouling potential and removal in UF systems for seawater RO pretreatment.
- To investigate the feasibility of coating UF membranes to stabilize UF operation, at low coagulant dose, on algal bloom-impacted water containing high AOM concentrations and to identify mechanisms through which coating may successfully mitigate MF/UF fouling during algal bloom periods.

- To measure AOM removal rates from RO feed water with conventional coagulation (coagulation/flocculation/sedimentation/filtration) in terms of organic surrogate parameters (biopolymers, TEP, UV_{254}, SUVA), inorganic surrogate parameters (Fe, turbidity) and MFI-UF.
- To evaluate the feasibility of low molecular weight cut-off UF membranes as a coagulant free alternative for SWRO pretreatment.

1.7 Outline of thesis

This thesis has been structured in nine chapters. **Chapter 1** is a general introduction on the background of the study - including operational problems associated with algal blooms in SWRO pretreatment systems - highlighting the need for further investigation of the role of coagulation in UF operation as SWRO pretreatment during algal blooms.

Chapter 2 is a review of state-of-the-art on the potential impact of algal blooms on seawater reverse osmosis plants and pretreatment strategies to mitigate membrane fouling.

Chapter 3 demonstrates through theoretical calculations, particle properties that are most relevant to MF/UF operation with respect to pressure development in one filtration cycle. In addition, the extent to which a coagulant contributes to the remaining fouling in MF/UF membranes is investigated.

Chapter 4 elaborates on optimization of coagulation in MF/UF systems treating fresh and saline surface water with varying forms and concentrations of organic matter, and highlights the potential mechanisms through which inline coagulation enhances MF/UF operation.

Chapter 5 studies the effect of coagulation conditions on fouling potential and fouling mechanisms of AOM in UF systems and the extent to which AOM is removed by a combination of coagulation and UF in comparison with AOM removal by conventional coagulation.

Chapter 6 evaluates the feasibility of coating UF membranes to control fouling at low coagulant dose during periods of severe algal blooms. The parameters that affect UF coating efficiency and the mechanisms with which coating enhances UF operation are discussed and verified experimentally.

Chapter 7 investigates AOM removal rates by conventional coagulation by measuring a number of surrogate parameters for the characterization of AOM in seawater.

Chapter 8 investigates the feasibility of low MWCO UF as a coagulant free alternative for SWRO pretreatment. Low MWCO UF (10 kDa) operation is compared with standard commercially available UF membranes (150 kDa) in terms of hydraulic performance and permeate quality.

Chapter 9 summarizes the results of this study and gives recommendations for practice and further research.

1.8 References

Alldredge, A.L., Passow, U., & Logan, B.E., 1993. The abundance and significance of a class of large, transparent organic particles in the ocean. Deep-Sea Research I 40, 1131-1140.

Anderson, D.M., Cembella, A.D. and Hallegraeff, G.M., 2012. Progress in understanding harmful algal blooms: paradigm shifts and new technologies for research, monitoring and management. Annual Review Marine Science 4, 143-176.

Basha, K.S.A., Gulamhusein, A.H., Khalil, A.A., 2011. Successfully operating a seawater UF pretreatment system. IDA Journal, Desalination and Water Reuse 3(1), 14-18.

Berman T., Holenberg, M., 2005. Don't fall foul of biofilm through high TEP levels. Filtration & Separation 42, 30-32.

Brehant, A., Bonnelye, V. & Perez, M., 2002. Comparison of MF/UF pretreatment with conventional filtration prior to RO membranes for surface seawater desalination. Desalination 144, 353-360.

Busch, M., Chu, R. & Rosenberg, S., 2010. Novel trends in dual membrane systems for seawater desalination: minimum primary pretreatment and low environmental impact treatment schemes. IDA Journal, Desalination and Water Reuse 2(1), 56-71.

Caron , D.A., Garneau, M.E., Seubert, E., Howard M.D.A., Darjany L., Schnetzer A., Cetinic, I., Filteau, G., Lauri, P., Jones, B., Trussell, S., 2010. Harmful algae and their potential impacts on desalination operations off southern California. Water Research 44 (2), 385-416.

DesalData, 2013. IDA Desalination Plants Inventory. In: Global Water Intelligence and Water Desalination Report (eds.).

Edzwald, J.K., 2010. Dissolved air flotation and me. Water Research 44 (7), 2077-2106.

Edzwald, J.K., Haarhoff, J., 2011. Seawater pretreatment for reverse osmosis: Chemistry, contaminants, and coagulation. Water Research 45, 5428-5440.

Falkowski P.G., Barber R.T., Smetacek V., 1998. Biogeochemical controls and feedbacks on ocean primary production. Science 281 (5374), 200-206.

Fogg, G.E., 1983. The ecological significance of extracellular products of phytoplankton photosynthesis. Bot. Mar. 26, 3-14.

Gallego, S. & Darton, E., 2007. Simple laboratory technique improve the operation of RO pretreatment systems. In International Desalination Association World Congress, Maspalomas, Gran Canaria.

Gille, D. & Czolkoss, W., 2005. Ultrafiltration with multi-bore membranes as seawater pretreatment. Desalination, 182 (1-3), 301-307.

Glueckstern, P., Priel, M. & Wilf, M., 2002. Field evaluation of capillary UF technology as a pretreatment for large seawater RO systems. Desalination 147, 55-62.

GWI, 2007. Desalination markets 2007, a global industry forecast, Global Water Intelligence, Media Analytics Ltd., Oxford, UK.

Halpern, D.F., McArdle, J. & Antrim, B., 2005. UF pretreatment for SWRO: pilot studies. Desalination 182, 323-332.

Hoof, S.C.J.M.V., Minnery, J.G. & Mack, B., 2001. Dead-end ultrafiltration as alternative pretreatment to reverse osmosis in seawater desalination®: a case study. Desalination 139, 161-168.

Kim, S.H., Lee, S.H., Yoon, J.S., Moon, S.Y., Yoon, C.H., 2007. Pilot plant demonstration of energy reduction for RO seawater desalination through a recovery increase. Desalination 203 (1-3), 153-159.

Kim, S.H., Yoon, J.H., 2005. Optimization of microfiltration for seawater suffering from red tide contamination. Desalination 182, 315-321.

Ladner, D.A., Vardon, D.R., Clark M.M., 2010. Effects of shear on microfiltration and ultrafiltration fouling by marine bloom-forming algae. Journal of Membrane Science 356, 33-43.

Latteman, S., Höpner, T., 2008. Environmental impact and impact assessment of seawater desalination. Desalination 220, 1-15.

Lattemann, S., 2010. Development of an environmental impact assessment and decision support system for seawater desalination plants. Doctoral dissertation, UNESCO-IHE/TU Delft, Delft.

Laycock, M.V., Anderson, D.M., Naar, J., Goodman, A., Easy, D.J., Donovan, M.A., Li, A., Quilliam, M.A., Al Jamali, E., Alshihi, R., Alshihi, R., 2012. Laboratory desalination experiments with some algal toxins. Desalination 293, 1-6.

Le Gallou, S., Bertrand, S., Madan, K.H., 2011. Full coagulation and dissolved air flotation: a SWRO key pretreatment step for heavy fouling seawater. In International Desalination Association World Congress, Perth, Australia.

Matin, A., Khan, Z., Zaidi, S.M.J., Boyce, M.C., 2011. Biofouling in reverse osmosis membranes for seawater desalination: Phenomena and prevention. Desalination 281, 1-16.

Mopper, K., Zhou, J., Sri Ramana, K., Passow, U., Dam, H.G., Drapeau, D.T., 1995. The role of surface-active carbohydrates in the flocculation of a diatom bloom in a mesocosm. Deep-Sea Research II 42(1), 47-73.

Myklestad, S.M., 1995. Release of extracellular products by phytoplankton with special emphasis on polysaccharides. Science of the Total Environment 165, 155-164.

Pearce, G., 2007b. Introduction to membranes: Manufacturers' comparison: part 1. Filtration & Separation, 36-38.

Pearce, G., 2009. SWRO pretreatment: treated water quality. Filtration & Separation 46 (6), 30-33.

Pearce, G.K., 2007a. The case for UF/MF pretreatment to RO in seawater applications. Desalination 203 (1-3), 286-295.

Pearce, G.K., 2011. The role of coagulation in membrane pretreatment for seawater desalination. In International Desalination Association World Congress, Perth, Australia.

Petry, M., Sanz, M.A., Langlais, C., Bonnelye, V., Durand, J.P., Guevara, D., Nardes, W.M., Saemi, C.H., 2007. The El Coloso (Chile) reverse osmosis plant. Desalination 203 (1-3), 141-152.

Qu, F., Liang, H., Tian, J., Yu, H., Chen, Z., Li, G., 2012. Ultrafiltration (UF) membrane fouling caused by cyanobateria: Fouling effects of cells and extracellular organics matter (EOM). Desalination 293, 30-37.

Richlen, M.L., Morton, S.L., Jamali, E.A., Rajan, A., Anderson, D.M., 2010. The catastrophic 2008-2009 red tide in the Arabian Gulf region, with observations on the identification and phylogeny of the fish-killing dinoflagellate Cochlodinium polykrikoides. Harmful Algae 9, 163-172.

Schippers, J.C., 2012. Workshop dissolved air flotation: a threat or a blessing? Enschede, The Netherlands.

Schlosser, C.A., Strzepek, K., Gao, X., Gueneau, A., Fant, C., Paltsev, S., Rasheed, B., Smith-Greico, T., Blanc, E., Jacoby, H., Reilly, J., 2014. The future of global water stress: an integrated assessment. MIT Joint Program on the Science and Policy of Global Change. Report No. 254.

Schurer, R., Janssen, A., Villacorte, L., Kennedy, M.D., 2012. Performance of ultrafiltration & coagulation in an UF-RO seawater desalination demonstration plant. Desalination & Water Treatment, 42(1-3), 57-64.

Schurer, R., Tabatabai, A., Villacorte, L., Schippers, J.C., Kennedy, M.D., 2013. Three years operational experience with ultrafiltration as SWRO pretreatment during algal bloom. Desalination & Water Treatment, 51 (4-6), 1034-1042.

Sellner, K.G.,Doucette, G.J., Kirkpatrick, G.J., 2003. Harmful algal blooms: causes, impacts and detection. Journal of Ind. Microbiol. Biotechnol 30, 383-406.

UNEP, 2014. Cities and coastal areas, United Nations Environmental Programme (UNEP). http://www.unep.org/urban_environment/issues/coastal_zones.asp

Villacorte, L.O., 2014. Algal blooms and membrane based desalination technology. Doctoral dissertation, UNESCO-IHE/TU Delft, Delft.

Voutchkov, N., 2009. SWRO pretreatment systems: Choosing between conventional and membrane filtration. Filtration and Separation 46 (1), 5-8.

WHO, 2007. Desalination for Safe Water Supply, Guidance for the Health and Environmental Aspects Applicable to Desalination, World Health Organization (WHO), Geneva.

Wilf, M. & Schierach, M.K., 2001. Improved performance and cost reduction of RO seawater systems using UF pretreatment. Desalination 135, 61-68.

Wolf, P.H., Siverns, S. & Monti, S., 2005. UF membranes for RO desalination pretreatment. Desalination 182, 293-300.

WWAP - World Water Assessment Programme, 2012. The United Nations World Water Development Report 4: Managing water under uncertainty and risk. Paris, UNESCO.

2

SEAWATER REVERSE OSMOSIS AND ALGAL BLOOMS

This chapter provides an overview of state-of-the-art pretreatment options for seawater reverse osmosis, the effect of algal blooms on membrane filtration systems (e.g., micro- and ultrafiltration and reverse osmosis membranes), indicators for quantifying these effects and promising pretreatment options for RO desalination of algal bloom-impacted seawater.

2.1 Background

Reverse osmosis (RO) is currently the leading and preferred seawater desalination technology (DesalData, 2013). The main drawback for cost-effective application of RO is membrane fouling (Flemming et al., 1997; Baker and Dudley, 1998). The accumulation of particulate and organic material from seawater and biological growth in membrane modules frequently cause operational problems in seawater reverse osmosis (SWRO) plants. To reduce the (in)organic load of colloidal and particulate matter reaching RO membranes and to minimize or delay associated operational problems, pretreatment systems are generally installed upstream of the RO membranes. Most SWRO plants, especially in the Middle East, install coagulation followed by granular media filtration (GMF) to pretreat seawater. However, in recent years, low pressure membrane filtration is increasingly being used as SWRO pretreatment.

Over the years, it is becoming more evident that microscopic algae are a major cause of operational problems in SWRO plants (Caron et al., 2010). The adverse effect of algae on SWRO started to gain more attention during a severe algal bloom incident in the Gulf of Oman in 2008-2009 (Figure 2-1). This bloom forced several SWRO plants in the region to reduce or shutdown operation due to clogging of pretreatment systems (mostly GMF) and/or due to low RO feed water quality (i.e., high silt density index, SDI > 5). The latter triggers concerns of irreversible fouling problems in RO membranes (Berktay, 2011; Richlen et al., 2010; Nazzal, 2009; Pankratz, 2008). This incident highlighted a major problem that algal blooms may cause in countries relying largely on SWRO plants for their water supply, and underlines the significance of adequate pretreatment in such systems. The high particle loading in seawater during an algal bloom in combination with high filtration rates (5-10 m/h) in the filters can cause rapid and irreversible clogging of the GMF. Furthermore, a substantial fraction of algal-derived organic matter (AOM) can pass through GMF, which can potentially cause fouling in downstream RO membranes (Guastalli et al., 2013).

In 2005, Berman and Holenberg reported that some AOM, particularly transparent exopolymer particles (TEP), can potentially initiate and enhance biofouling in RO systems (Berman and Holenberg, 2005). TEP are characteristically sticky, so they can adhere and accumulate on the surface of RO membranes and spacers. The accumulated TEP may serve as a "conditioning layer" - a platform for effective attachment and initial colonization by bacteria - which may then accelerate biofilm formation in RO membranes (Bar-Zeev et al., 2012). Furthermore, TEP may be partially degradable and may later serve as a substrate for bacterial growth (Passow and Alldredge, 1994; Alldredge et al., 1993).

To solve the problem of high AOM concentration in the GMF filtrate, a few options have been proposed such as incorporating and/or increasing coagulant dosage prior to GMF to improve effluent water quality. However, an increase in coagulant dosage may further increase the rate of filter clogging. The addition of coagulant at high dose results in the creation of large flocs that are captured on the surface of the filters rather than being filtered through the media. This shifts the filtration mechanism from depth filtration to surface blocking, which at filtration rates of typically 5 - 10 m/h gives rise to significant head loss in these systems. Installing a floc removal step such as sedimentation or dissolved air flotation (DAF) in front of the GMF will reduce clogging problems in GMF when high coagulant dose is required to improve filtrate quality.

Figure 2-1 A massive red tide bloom in the Gulf of Oman spreading to the Persian Gulf shown in this satellite image acquired by Envisat's MERIS instrument on November 22, 2008 (Credits: C-wams project, Planetek Hellas/ESA). Yellow points indicate locations of major SWRO plants in the area. Inset screenshots of online news on SWRO plant shutdown due to red tide in the region in 2008 and 2013 (www.arabianbusiness.com)

Another option is to install ultrafiltration (UF) or microfiltration (MF) membrane systems to replace GMF. UF/MF pretreatment can guarantee an RO feed water with low SDI even during severe algal bloom. However, concerns have been expressed regarding the rate of fouling in UF/MF membrane systems (e.g., backwashable and non-backwashable fouling) during algal bloom periods (Schurer et al., 2013). To overcome this concern, DAF preceding MF/UF has been recommended (Anderson and McCarthy, 2012).

This chapter reviews state-of-the-art pretreatment options for SWRO, the effect of algal blooms on membrane filtration systems (e.g., MF/UF and RO), indicators for quantifying these effects and promising pretreatment options for RO desalination of algal bloom-impacted seawater.

2.2 Seawater pretreatment options

Most SWRO plants are equipped with one or more pretreatment systems. These mainly include granular media filtration (GMF), dissolved air flotation (DAF) and/or ultrafiltration (UF).

2.2.1 Granular media filtration (GMF)

Conventional pretreatment systems for SWRO were developed based on existing technology and most commonly consist of conventional media filtration. Single or dual stage granular media filters consisting of sand and anthracite (garnet is sometimes used) is typically applied in conventional pretreatment systems, in gravity or pressurized configuration. Sand and anthracite (0.8-1.2 mm/2-3 mm) filter beds are superior to single media filtration in that they provide higher filtration rates, longer runs and require less backwash water. Anthracite/sand/garnet beds have operated at normal rates of approximately 12 m/h and peak rates as high as 20 m/h without loss of effluent quality. In SWRO pretreatment, the primary function of GMF is to reduce high loads of particulate and colloidal matter (i.e., turbidity).

GMF relies on depth filtration to enhance RO feed water quality. However, when high concentrations of organic matter or turbidity loads are encountered, coagulation is required to ensure that RO feed water of acceptable quality is produced (SDI < 5). Coagulation is applied either in full scale or inline mode in these systems. The most commonly applied coagulant in SWRO pretreatment is ferric salts (i.e., ferric chloride or ferric sulphate).

Poor removal of algae can lead to clogging of granular media filters and short filter runs. While diatoms are well-known filter clogging algae, other algae types can clog filters including green algae, flagellates, and cyanobacteria (Edzwald, 2010). During the severe red tide bloom event in the Gulf of Oman and Persian Gulf in 2008-09, conventional pretreatment systems were not able to maintain production capacity at high algal cell concentrations of approximately 27,000 cells/mL (Richlen et al., 2010). Operation of the media filters was characterized by rapid clogging rates and deteriorating quality of pre-treated water. As a consequence, frequent backwashing was required resulting in increased downtime of the system such that the required pretreatment capacity could no longer be maintained.

At the Fujairah plant in UAE, filter runs were reduced from 24 to 2 hours. Furthermore, deteriorating quality of the pre-treated water, i.e., SDI > 5, led to higher coagulant dose required to enhance treated water quality. Increasing coagulant dose may lead to higher clogging rates of media filters. Coagulation enlarges particulate and colloidal matter in water and can therefore shift filtration mechanism from standard blocking (depth filtration) to surface blocking (cake filtration). As filtration rates are relatively high (5-10 m/h) in media filters, cake filtration can result in exponential head loss in the filters.

Reducing filtration rates of the media filters during such extreme events can improve operation. Reducing the rate of filtration by 50% will result in much lower clogging rates, e.g., by a factor 2-4 depending on the size and characteristics of the foulants (e.g., algae). However, reducing filtration rates will require increased surface area of the media filters. This implies significant investment costs and larger foot print of the treatment plant. Another way to enhance operation of GMF during such extreme events is to provide a clarification step, e.g., sedimentation or flotation, after coagulation/flocculation to reduce the load of particulate/colloidal matter (including coagulated flocs) on the media filters.

Flotation is more robust than sedimentation as it can handle large concentrations of suspended matter (e.g., algae). Currently, flotation preceding media filtration is proposed as the solution for algal blooms. Flotation is able to reduce the algal concentration to a large extent, protecting media filters from rapid clogging, reduced capacity, and breakthrough. However a coagulant dose of 1-2 mg Fe(III)/L or higher is usually required to render the process effective. Furthermore, coagulant might be required upstream of the media filters to ensure an acceptable SDI in the effluent. Installing flotation units in front of media filtration might be cheaper than sedimentation units, as the surface loading rates in high rate DAF systems can reach 30 m/h (Edzwald, 2010). Consequently, flotation may require much lower footprint than sedimentation. However the process scheme will require flocculation basins, air saturation and sludge treatment facilities.

2.2.2 Dissolved air flotation (DAF)

DAF is a clarification process that can be used to remove particles prior to conventional media filtration or MF/UF systems. Raw water is dosed with a coagulant, typically at concentrations lower than those applied for sedimentation, followed by two-stage tapered flocculation. Removal is achieved by injecting the feed water stream with water that has been saturated with air under pressure and then releasing the air at atmospheric pressure in a flotation tank. As the pressurized water is released, a large number of micro-bubbles are formed (approximately 30-100 μm) that adhere to coagulated flocs and suspended matter causing them to float to the water surface where they may be removed by either a mechanical scraper or hydraulic means, or a combination thereof. Clarified water (sub-natant) is drawn off the bottom of the tank by a series of lateral draw-off pipes (Figure 2-2). Conventional DAF systems operate at nominal hydraulic loading rates of 5-15 m/h. More recent DAF units are developed for loadings of 15-30 m/h and greater. As a result, DAF requires smaller footprint than sedimentation.

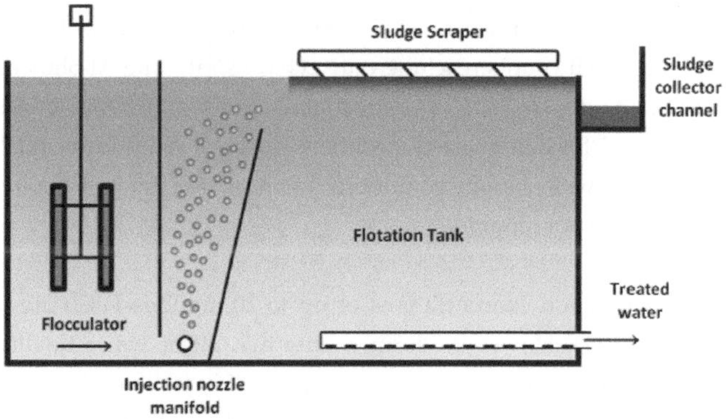

Figure 2-2 Schematic presentation of a simplified DAF unit with flocculator

DAF is more effective than sedimentation in removing low density particles from water and is therefore a suitable treatment process for algal bloom-impacted waters. Gregory and Edzwald (2010) reported 90-99% removal by DAF of algal cells for different algae types compared to 60-90% by sedimentation. A review paper on separation of algae by Henderson et al. (2008) reports DAF removals of 96% to about 99.9% when pretreatment and DAF are optimized. Several DAF plants in the Netherlands and Great Britain are primarily used for treatment of algal-laden waters (van Puffelen et al., 1995; Longhurst and Graham, 1987; Gregory, 1997).

In SWRO pretreatment, DAF prior to dual-stage GMF was tested by Degrémont during early pilot testing for the Taweelah SWRO plant in Abu Dhabi, UAE (Rovel, 2003). DAF was suggested to enhance the robustness of the pretreatment scheme in case of oil spills or algal bloom events, or in case high coagulant concentrations were required during turbidity spikes. Algal cell concentrations were reportedly below 100 cells/mL during this period, which is far below concentrations observed during severe bloom conditions. Sanz et al. (2005) demonstrated the effectiveness of DAF coupled with coagulation prior to dual stage GMF in producing RO feed water with SDI < 4 (typically less than 3) when treating seawater containing various algae, including red tide species. The paper does not specify the initial total cell concentration of various species. The authors reported more than 99% removal of total algae after DAF and first stage filtration units.

The severe red tide event in 2008-2009 in the Gulf of Oman that led to the shutdown of several desalination plants in the region redirected the attention of the desalination industry to DAF as part of SWRO pretreatment schemes. DAF is now being regularly used in new SWRO plants in the Persian Gulf upstream of granular media filtration or UF. Although the Fujairah 2 desalination plant was still under construction during that period, Veolia reported that their pilot plant fitted with a DAF unit in the pretreatment system, continued to operate throughout the red tide bloom (Pankratz, 2008). Expansion of the Fujairah plant incorporates DAF as an essential part of the pretreatment scheme (WaterWorld, 2013). Degrémont reported > 99% removal of algal cells during pilot testing of coagulation/AquaDAF™ prior to GMF in Al-Dur, Bahrain (Le Gallou et al., 2011). However, real bloom conditions were not encountered during the pilot phase with algal cell counts reaching only 200 cells/mL. The Al-Shuwaikh desalination plant in Kuwait equipped with DAF/UF as pretreatment consistently provided SDI < 2.5 for good quality feed water and <3.5 for deteriorated conditions during a red tide event (Park et al., 2013). However, in this case as well, bloom conditions as measured by cell counts, chlorophyll-a concentrations, or TEP were not reported.

In most DAF units, coagulation concentrations of up to 20 mg/L as $FeCl_3$ are reported (Rovel, 2003; Le Gallou et al., 2011). However, storm events that affect water quality may result in substantially higher coagulant concentrations for DAF (Le Gallou et al., 2011). Moreover, additional coagulant dosage is often required in GMF units downstream of DAF.

2.2.3 Ultrafiltration (UF)

Over the last decade, the application of UF has been considered as a more reliable alternative to conventional granular media filtration (with and without coagulation) as a pretreatment process for RO systems. UF membranes have been tested and applied at pilot and commercial scale as pretreatment for SWRO (Wolf et al., 2005; Halpern et al., 2005; Gille and Czolkoss, 2005; Brehant et al., 2002; Glueckstern et al., 2002; Wilf and Schierach, 2001) and offer several advantages over conventional pretreatment systems; namely, lower footprint, constant high permeate quality (in terms of SDI), higher retention of large molecular weight organics, lower overall chemical consumption, etc. (Wilf and Schierach, 2001; Pearce, 2007). Successful piloting has led to the implementation of UF pretreatment in several large (> 100,000 m^3/day) SWRO plants (Busch et al., 2010), with a total installed capacity of approximately 5 million m^3/day as of 2013.

UF membranes are generally more effective in removing particulate and colloidal matter from seawater than GMF. Hence, they are expected to be more reliable in maintaining an RO feed water with low fouling potential even during an algal bloom period. However, MF/UF membranes were also reported to experience some degree of fouling during algal blooms (Schurer et al., 2012; 2013). So far, a few studies have investigated the effect of algal blooms on UF membrane operation (Kim and Yoon, 2005; Ladner et al., 2010, Schurer et al., 2012; 2013). These studies agree on the notion that large macromolecules (e.g., polysaccharides and proteins) produced by these algae are the main causes of membrane fouling, and more so than the algal cell themselves. High concentrations of sticky AOM substances (e.g., TEP) present during an algal bloom can impair UF operation (Figure 2-3) by attaching to the membrane surface and pores resulting in permeability decline (CEBs as frequent as once in 6 hours).

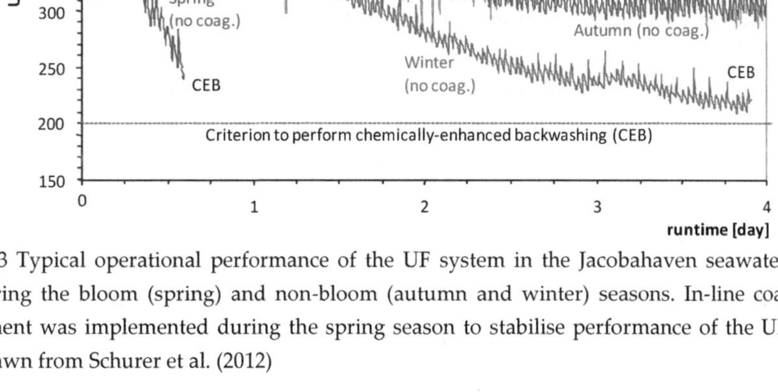

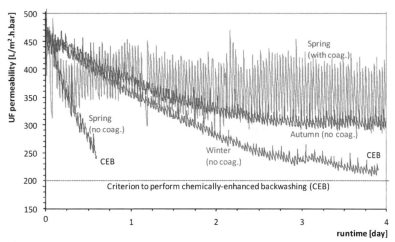

Figure 2-3 Typical operational performance of the UF system in the Jacobahaven seawater UF-RO plant during the bloom (spring) and non-bloom (autumn and winter) seasons. In-line coagulation pretreatment was implemented during the spring season to stabilise performance of the UF. Graph was redrawn from Schurer et al. (2012)

Under such conditions, operators resorted to coagulation to stabilize operation (Schurer et al., 2012; 2013). With optimized coagulation conditions, operation was stabilized at relatively low doses of ferric during the bloom period. Although extensive operational experience on algal blooms of different types and severities is not available, one can extrapolate the results of the existing pilot study and propose that inside-out pressure driven UF membranes are more capable of handling algal bloom events than conventional GMF. This may be attributed to significant differences in hydraulic operational parameters of the two systems.

An indicative overview of operational parameters for media filtration and UF is presented in Table 2-1. Filtration flux rates in GMF can be up to 100 times higher than flux rates in UF systems. Total filtered volume prior to backwash may be 2000 times higher for GMF compared with UF membranes. Hence, low filtration rates and high backwash frequencies favour overall enhancement of UF systems performance. However, coagulation is required to stabilize hydraulic performance during periods of severe algal bloom.

Table 2-1 Operational parameters for ultrafiltration and media filtration typically applied in SWRO pretreatment (adapted from Schippers, 2012)

	Ultrafiltration	Granular media filtration
Pores [μm]	0.02	150
Filtration rate [L/m²h]	50 – 100	5,000 – 10,000
Run length [h]	1	24
Ratio backwash rate : filtration rate	2.5	2.5 - 5
Backwash time [min]	1	30
Filtered volume/m² per cycle [L]	50 – 100	120,000 – 240,000
Pressure loss [bar]	0.2 – 2	0.2 – 2

Coagulation is commonly applied in inline mode in UF systems for SWRO pretreatment. Inline coagulation is the application of a coagulant without removal of coagulated flocs through a clarification step. Inline coagulation may also be characterized by the absence of a flocculation chamber. Hence, in most inline coagulation applications, coagulation is achieved by dosing the coagulant prior to a static mixer or the feed pump of UF membranes to provide adequate mixing. Flocculation is generally not required in UF applications, as enlarging particle size is not an objective and pin-sized flocs are sufficient to enhance UF operation.

However, if not optimized, coagulation can deteriorate long-term UF operation. Ferric species (monomers, dimers, trimers, etc.) that are small enough to enter UF pores, can irreversibly foul UF membranes. If low grade coagulants are used, ferrous iron can reach UF membranes and adsorb on the membrane surface or within the pores. Fouling by manganese may also occur for coagulants of low grade. This may result in slow irreversible fouling of UF membranes that can only be removed with specialized cleaning solutions with proprietary recipes. Some chemicals that can help reduce irreversible fouling of UF membranes by iron and manganese are solutions based on ascorbic and oxalic acids of a certain ratio.

2.3 Impact of algal blooms on UF and SWRO

Caron et al. (2010) pointed out two potential impacts of algal blooms on membrane-based seawater desalination facilities,

1) Significant treatment challenge to ensure the desalination systems are effectively removing algal toxins from seawater;
2) Operational difficulties due to increased total suspended solids and organic content resulting from algal biomass in the raw water.

The presence of algal toxins during marine algal blooms is a growing concern, since some of these toxins are highly toxic (Anderson and McCarthy, 2012). Studies on harmful algal bloom (HAB) toxins removal by reverse osmosis are scarce and limited to laboratory bench-scale studies. These studies suggest that 99% removal can be achieved with RO membranes (Laycock et al., 2012). Whether this level of rejection is adequate cannot be justified as data from operational plants during toxic HABs is not available and there are no WHO regulations for marine HAB toxins in drinking water.

From an operational point of view, algal blooms can pose a real concern for SWRO plants with conventional (GMF) or advanced (MF/UF) pretreatment systems.

2.3.1 Particulate fouling

High algae biomass in raw water can cause operational problems in membrane systems. During filtration of algal bloom-impacted waters, particulate matter comprising algal cells, their detritus and AOM can accumulate to form a heterogeneous and compressible cake layer on the surface of the membranes which may eventually cause a substantial decrease in overall membrane permeability.

RO/NF systems are primarily designed to remove dissolved constituents in the water but they are most vulnerable to spacer clogging problems by particulate matter. For this reason, the majority of RO systems are preceded by a pretreatment process to minimise particulate fouling potential of the feed water. Nevertheless, common pretreatment processes such as granular (dual) media filters may not be reliable to prevent particulate fouling during algal bloom (Berktay, 2011; Nazzal, 2009; Anderson and McCarthy, 2012). The product water of granular media filters can be highly variable over time, with reported algae and biopolymer (algal-released organic macromolecules) removal efficiencies in the range of 48-90% and 17-47%, respectively (Plantier et al., 2012; Salinas Rodriguez et al., 2009).

Capillary MF/UF membranes - operating in inside-out mode - may potentially experience different ways of fouling when very high numbers of algal cells are present in the feed water, e.g.,

- Uniform deposition of algal cells along the membrane capillary;

- Deposition primarily at the capillary entrance;
- Deposition primarily at the capillary dead-end.

Modelling work by Panglisch (2003) and Lerch (2008) indicate that small particles tend to deposit uniformly along the capillary length whereas larger particles tend to deposit primarily near the capillary dead-end (see Figure 2-4). This indicates that during algal blooms and in the presence of large numbers of algal cells, a part of the capillary (at the dead-end) might be filled with algal cells and associated algal organic matter. This might result in a higher flux at the capillary inlet, assuming that the parts of the capillary with algae and organic matter deposition are no longer permeable. Villacorte (2014) demonstrated that, even at very high numbers, algal cell deposition does not severely limit membrane permeability.

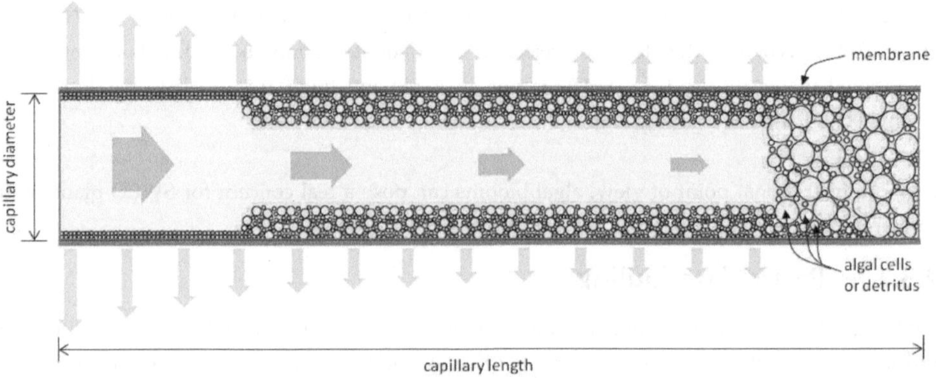

Figure 2-4 Graphical illustration of capillary UF membrane plugging due to accumulation of algal cells and their detritus (Panglisch, 2003)

Heijman et al. (2005; 2007) demonstrated that larger particles might clog the entrance of capillaries. This phenomenon might be avoided by applying micro-screens with openings smaller 150-300 μm which is currently applied.

2.3.2 Organic fouling

During algal blooms, organic fouling in UF membranes often occurs when AOM are abundant in the feed water. AOM produced by common bloom-forming species of algae largely comprise high molecular weight biopolymers (polysaccharides and proteins) which often include sticky TEP (Myklestad, 1995; Villacorte et al., 2013). TEP are hydrophilic materials that can absorb/retain water up to ~ 99% of their dry weight while allowing some water to pass through (Verdugo et al., 2004; Azetsu-Scott and Passow, 2004). This means that they can bulk-up to more than 100 times their solid volume and can squeeze through and fill-up the interstitial voids between the accumulated solid particles (e.g., algal cells) on the surface of the membrane (Figure 3-5). It is therefore expected that accumulation of these materials can provide substantial resistance to permeate flow during membrane filtration. As TEP can be very sticky, it may strongly adhere to the surface and pores of UF membranes. Consequently, hydraulic cleaning (backwashing) may no longer be effective in adequately restoring initial membrane permeability (Figure 2-5). This scenario has been reported in recent studies (e.g., Villacorte et al., 2010a,b; Schurer et al., 2012; 2013; Qu et al., 2012a,b), signifying that AOM could not only cause pressure increase during filtration but may also increase non-backwashable or physically irreversible fouling in dead-end UF systems.

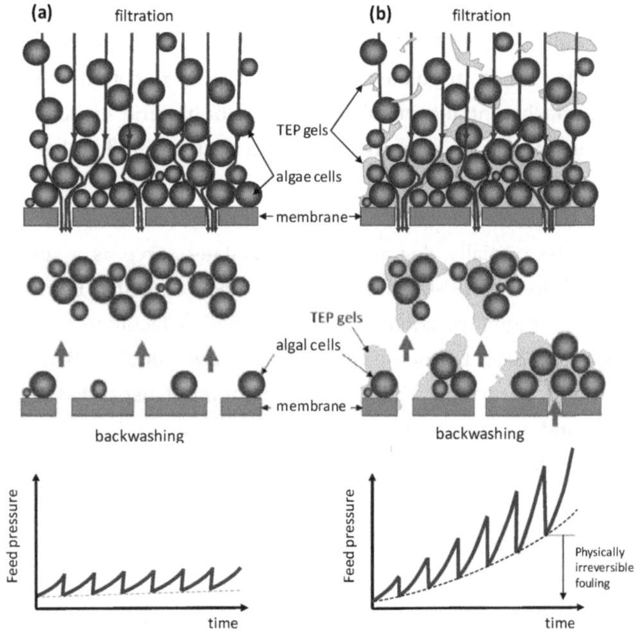

Figure 2-5 Graphical presentation of the potential role of TEP in fouling of UF membranes during severe algal blooms; filtration during algal bloom with (a) low and (b) high concentrations of TEP

Accumulation of AOM on RO membranes may result in lower normalized flux and higher feed channel pressure drop. Considering the high operating pressure used in RO, the direct impact of TEP accumulation on operational performance is expected to be much less remarkable than in UF systems. However, the accumulated sticky substances may initiate or promote particulate and biological fouling by enhancing deposition of bacteria and other particles from the feed water to the membrane and spacers (Winters and Isquith, 1979; Berman and Holenberg, 2005).

2.3.3 Biological fouling

Bacteria has been shown to adhere, accumulate and multiply in RO systems which eventually result in the formation of a slimy layer of dense concentrations of bacteria and their extracellular polymeric substances known as biofilm. When the accumulation of biofilm reaches a certain threshold that operational problems are encountered in the membrane system, it is considered as biological fouling or biofouling (Flemming, 2002). An operational problem threshold can be a remarkable (e.g., > 15%) decrease of normalised membrane flux, increase in net driving pressure and/or increase in feed channel pressure drop. In the Middle East, about 70% of the seawater RO installations were reported to be suffering from biofouling problems (Gamal Khedr, 2000). Generally, biofouling is only a major problem in NF/RO systems because periodic backwashing and chemical cleaning in dead-end UF systems allows regular dispersion or removal of most of the accumulated bacteria from the membrane; thus, inhibiting the formation of a biofilm.

TEP produced during algal blooms can initiate and enhance biofouling in RO systems. Because TEP are characteristically sticky, they can adhere and accumulate on the surface of the membranes and spacers. The accumulated TEP can serve as a "conditioning layer" – a good platform for effective attachment and initial colonization of bacteria - where bacteria can utilize effectively biodegradable nutrients from the feed water (Berman and Holenberg, 2005; Winters and Isquith, 1979).

Furthermore, TEP can be partially degraded and may later serve as a substrate for bacteria growth (Passow and Alldredge, 1994; Alldredge et al., 1993). Recently, Berman and co-workers proposed a "revised paradigm" of aquatic biofilm formation facilitated by TEP (Berman and Holenberg, 2005; Berman, 2010; Berman et al., 2011; Bar-Zeev et al., 2012a). As illustrated in Figure 2-6, colloidal and particulate TEP and protobiofilms (suspended TEP with extensive microbial outgrowth and colonization) in surface water can initiate, enhance and possibly accelerate biofilm accumulation in RO membranes.

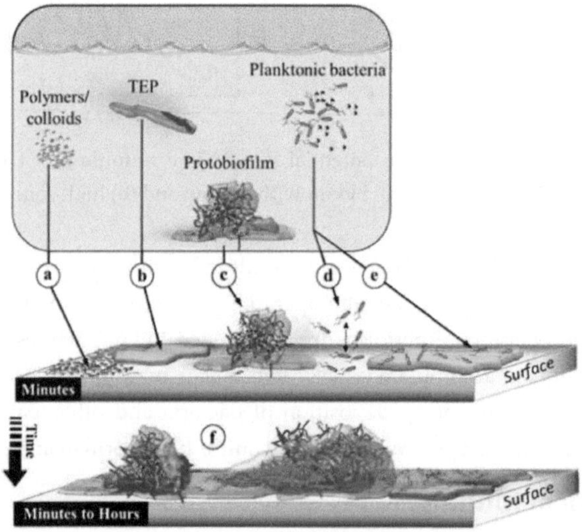

Figure 2-6 Schematic illustration of the possible involvement of (a) colloidal biopolymers, (b) TEP, and (c) protobiofilm in the initiation of aquatic biofilms. A number of planktonic bacteria (first colonizers) can attach (d) reversibly on clean surfaces or (e) irreversibly on TEP-conditioned surfaces. When nutrients are not limited in the water, (f) a contiguous coverage of mature biofilm can develop within a short period of time (minutes to hours). Figure and description Bar-Zeev et al. (2012a)

Since bacteria require nutrients for energy generation and cellular biosynthesis, essential nutrients such as biodegradable organic carbon (BDOC), phosphates and nitrates can be the main factors dictating the formation and growth of biofilm. During the peak of an algal bloom, some of these essential nutrients can be limited (e.g., phosphate) due to algal uptake. However when the bloom reaches death phase, algal cells starts to disintegrate and release some of these nutrients. Hence, biofouling initiated/enhanced by TEP and AOM may occur within a period of time after the termination of an algal bloom.

2.4 Fouling potential indicators

Monitoring membrane fouling potential of raw and pretreated water is important in SWRO plants, especially during algal bloom periods, in order to develop preventive/corrective measures for minimizing the potential adverse impacts to membrane filtration. Various indicators have been proposed to assess the magnitude of the bloom and effectiveness of the pretreatment systems. The most relevant indicators/parameters are discussed in the following sections.

2.4.1 Algae concentration

The magnitude of algal blooms is measured either in terms of cell count or chlorophyll-a concentration. Ideally, the pretreatment systems of an SWRO plant should effectively remove algal cells to prevent clogging in RO channels. Table 2-2 shows removal efficiencies of selected pretreatment processes reported in literature. Bloom-forming algae of different species can vary substantially in terms of cell size and chlorophyll-a content. Hence, relationship between these two parameters also varies. Typical bloom concentrations are higher for smaller algae compared with larger algae (Villacorte, 2014).

Table 2-2 Reported removal efficiencies of algae for various treatment processes

Treatment	Source water	Remarks	Removal (%)	Ref.
Granular media filtration	E. Mediterranean Sea	Rapid sand filter (no coag.)	76 ± 13	[1]
	E. Mediterranean Sea	Coag. + mixed bed filter	90 ± 8	[2]
	E. Mediterranean Sea	Coag. [1 mg $Fe_2(SO_4)_3$] + RSF	79 ± 8	[3]
	W. Mediterranean Sea	Press. GMF (anthracite-sand)	74	[4]
	Algae-spiked seawater	Dual media filter (no coag.)	48 - 90	[5]
Sedimentation	Lake water	Coag. = 20-24 mg Fe/L	96	[6]
	algae-spiked freshwater	Coag. = 12 mg Al_2O_3/L	90	[7]
Dissolved air flotation (DAF)	W. Mediterranean Sea	Coag. (0-6 mg $FeCl_3$/L) + DAF	75	[4]
	Lake water	Coag. = 7-12 mg Fe/L	96	[6]
	Algae-spiked freshwater	Coag. = 12 mg Al_2O_3/L	96	[7]
	Algae-spiked freshwater	Coag. = 0.5-4 mg Al_2O_3/L	90 - 100	[8]
	Algal culture	Coag. = 0.7-3 mg Al/L	98	[9]
Microfiltration	Algal cultures	No coagulation	> 99	[10]
Ultrafiltration	W. Mediterranean Sea	PVDF; pore size = 0.02 μm	99	[4]
	Algae-spiked freshwater	PVC; nom. pore size = 0.01 μm	100	[11]

Table 2-2 Cont'd. Reported removal efficiencies of algae for various treatment processes

Cartridge filters	E. Mediterranean Sea	Disruptor® media	60	[12]
	Lake Kinneret	Amiad™ AMF; 2-20 μm	90 ± 6	[13]
	Lake Kinneret	Disruptor® media	65	[12]
	River Jordan	Disruptor® media	85	[12]
	Treated wastewater	Disruptor® media	70	[12]

Note: Removal efficiency calculated based on chlorophyll-a concentration or cell count.

References: [1] Bar-Zeev et al. (2012b); [2] Bar-Zeev et al. (2009); [3] Bar-Zeev et al. (2013); [4] Guastalli et al. (2013) [5] Plantier et al. (2012); [6] Vlaski (1997) ; [7] Teixeira & Rosa (2007) ; [8] Teixeira et al. (2010); [9] Henderson et al. (2009) ; [10] Castaing et al. (2011) ; [11] Zhang et al. (2011) ; [12] Komlenic et al. (2013) ; [13] Eschel et al. (2013).

To compensate for the size differences, cell concentration can be expressed in terms of volume fraction (total cell volume per volume of water sample) instead of cell number per volume of water. Algae removal in granular media filters (GMF) is highly variable (48-90%) as compared to more stable and much higher removal efficiencies by MF/UF membranes (> 99%). High algal removals (> 75%) were also reported for sedimentation and DAF treatments. Cartridge filters, which are typically installed after the pretreatment processes and before the SWRO system, have comparable removal with GMF.

Although algal cell and chl-a concentrations are the main indicators of an algal bloom, these parameters are often not sufficient indicators of the fouling potential of the water. Different bloom-forming species of algae can behave differently in terms of AOM production and at which stage of their life cycle AOM materials are released. More advanced parameters that better indicate the concentration of AOM in feed water and the fouling potential attributed to the presence of AOM are discussed in the following sections.

2.4.2 Biopolymer concentration

Liquid chromatography-organic carbon detection (LC-OCD) is a semi-quantitative method for the measurement and fractionation of organic carbon. This method can be used to quantify the presence of AOM in algal bloom-impacted waters. Using this technique, AOM can be fractionated based on molecular size.

The high molecular weight fraction is classified as biopolymers while the low molecular weight fractions (< 1 kDa) are further sub-classified into humic-like substances, building blocks, acids and neutrals (Huber et al., 2011). Considering that the high molecular weight AOM are likely to deposit/accumulate in the RO system, measuring the biopolymer fraction of organic matter in the water is a promising indicator of organic and biological fouling potential of algal bloom impacted waters.

Reported biopolymer removal efficiencies of different treatment processes are presented in Table 2-3. Biopolymers in the water can be substantially reduced (> 50%) by UF and sub-surface intake (e.g., beach well) treatments while granular media filtration typically remove less than 50% of biopolymers.

Table 2-3 Reported removal efficiencies of various treatment processes based on biopolymer concentrations measured using LC-OCD

Treatment	Source water	Remarks	Removal (%)	Ref.
Granular media filtration	W. Mediterranean Sea	Coag. + dual media filter	47	[1]
	W. Mediterranean Sea	Press. GMF (anthracite-sand)	18	[2]
	E. Mediterranean Sea	Coag. + single media filter	32	[1]
	Estuarine (brackish)	Coag. + continuous sand filter	17	[1,3]
Microfiltration	W. Mediterranean Sea	PVDF; nom. pore size= 0.1 μm	36	[1]
	W. Mediterranean Sea	PVDF	14	[4]
	Red Sea	Ceramic; pore size = 0.08 μm	20	[5]
	Oman Gulf	Ceramic; pore size = 0.08 μm	30	[5]
	Algal cultures	PC; nom. pore size = 0.1 μm	52 - 56	[6]
Ultrafiltration	North Sea	MWCO = 300 kDa	50	[4]
	Red Sea	Ceramic; pore size = 0.03 μm	40	[5]
	W. Mediterranean Sea	PVDF; pore size = 0.02 μm	41	[2]
	Seawater (Sydney)	MWCO = 17.5 kDa	81.3	[7]
	Estuarine (brackish)	No coagulation	70	[1,3]
	River water	PVDF; pore size = 0.02 μm	86	[8]
	River water	PVDF; pore size = 0.01 μm	59	[9]
	Algal cultures	PES; MWCO = 100 kDa	65 - 83	[6]
	Algal cultures	RC; MWCO = 10 kDa	83 - 95	[6]
	Algae-spiked seawater	Ceramic; pore size = 0.03 μm	60	[5]
Subsurface intake	W. Mediterranean Sea	Beach well	70	[1]
	N. Pacific ocean	Infiltration gallery	75	[4]

References: [1] Salinas-Rodriguez et al. (2009); [2] Guastalli et al. (2013); [3] Villacorte et al. (2009a); [4] Salinas-Rodriguez (2011); [5] Dramas & Croué (2013); [6] Villacorte et al. (2013); [7] Naidu et al. (2013) [8] Hallé et al. (2009); [9] Huang et al. (2011).

2.4.3 Transparent exopolymer particles (TEP)

Transparent exopolymer particles (TEP) are a major component of the high molecular weight fraction (biopolymers) of algal organic matter. As discussed in Section 2.3, these materials can potentially cause severe fouling in UF and RO systems.

Over the last two decades, various methods were developed to measure TEP by microscopic enumeration or by spectrophotometric measurements (Alldredge et al., 1993; Passow and Alldredge, 1995; Arruda-Fatibello et al., 2004; Thornton et al., 2007). The most widely used and accepted method was introduced in 1995 by Passow and Alldredge. This method is based on retention of TEP on 0.4 μm polycarbonate membrane filters and subsequent staining with Alcian blue dye. The reported $TEP_{0.4\mu m}$ reduction by different pretreatment processes are summarised in Table 2-4.

Table 2-4 Reported removal efficiencies of $TEP_{0.4}$ for various treatment processes

Treatment	Source water	Remarks	Removal (%)	Ref.
Granular media filtration	E. Mediterranean Sea	Rapid sand filter (no coag.)	51 ± 27	[1]
	E. Mediterranean Sea	Coag. + mixed bed filter	27 ± 19	[2]
	E. Mediterranean Sea	Coag. [1 mg $Fe_2(SO_4)_3$] + RSF	17 ± 28	[3]
	Estuarine (brackish)	Coag. + continuous sand filter	65	[4,5]
	River water	Coag. (8 ml/L PACl) + RSF	25	[5]
	River water	Rapid sand filter (no coag.)	~100	[6]
	Treated wastewater	Coag. (10 mg Al/L) + RSF	70	[7]
Sedimentation + media filtration	Lake water	Coag. = 15 mg Fe/L	70	[5]
Ultrafiltration	Estuarine (brackish)	No coagulation	100	[4,5]
	Lake water	No coagulation	100	[5]
	River water	Inline coag. = 3 mg Fe/L	100	[5]
	River water	Inline coag. = 0.3 mg Fe/L	100	[5,8]
	Treated wastewater	No coagulation	100	[7]
	Treated wastewater	No coagulation	> 95	[6]
Cartridge filter	E. Mediterranean Sea	Disruptor® media	59	[9]
	Lake Kinneret	Amiad™ AMF; 2-20μm	47 ± 21	[10]
	Lake Kinneret	Disruptor® media	63	[9]
	River Jordan	Disruptor® media	82	[9]
	Treated wastewater	Disruptor® media	74	[9]

Note: $TEP_{0.4}$ measurement based on Passow and Alldredge (1995).

References: [1] Bar-Zeev et al. (2012b); [2] Bar-Zeev et al. (2009); [3] Bar-Zeev et al. (2013); [4] Salinas-Rodriguez et al. (2009); [5] Villacorte et al. (2009a); [6] van Nevel et al. (2012); [7] Kennedy et al. (2009); [8] Villacorte et al. (2009b); [9] Komlenic et al. (2013) ; [10] Eschel et al. (2013).

Although they are operationally defined by marine biologists as particles larger than 0.4 μm, TEP are not solid particles, but rather agglomeration of particulate and colloidal hydrogels which can vary in size from few nanometres to hundreds of micrometres (Passow, 2000; Verdugo et al., 2004). Hydrogels are highly hydrated and may contain more than 99% of water, which means they can bulk-up to more than 100 times their solid volume (Azetsu-Scott and Passow, 2004; Verdugo et al., 2004). Majority of these materials are formed abiotically through spontaneous assembly of colloidal polymers in water (Chin et al., 1998, Passow, 2000). A substantial fraction of colloidal TEP (< 0.4μm) are not covered by the current established measurement methods (e.g., Passow and Alldredge, 1995). Consequently, new TEP methods were recently developed to cover this previously neglected smaller fraction (see Villacorte, 2014).

2.4.4 Modified Fouling Index (MFI)

There are two established methods to measure the particulate/colloidal fouling potential of water; silt density index (SDI) and Modified Fouling Index (MFI). Currently, the silt density index (SDI) is the most widely used method to measure the fouling potential of feed water in SWRO plants. It is based on measurements using membrane filters with 0.45 micrometer pores at a pressure of 210 kPa (30 psi). Although this simple technique is widely used in practice, it has been known for many years that SDI has no reliable correlation with the concentration of particulate/colloidal matter (Alhadidi et al., 2013). Hence, it is often insufficient in predicting the fouling potential of SWRO feed water.

A more reliable approach to measure the membrane fouling potential of RO feed water is the modified fouling index (MFI). Unlike SDI, MFI is based on a known membrane fouling mechanism (i.e., cake filtration). This index was developed by Schippers and Verdouw (1980) whereby they demonstrated the linear correlation between the MFI and colloidal matter concentration in the water. Initially, MFI was measured using membranes with 0.45 or 0.05 μm pore sizes and at constant pressure. However, it was later found that particles smaller than the pore size of these membranes most likely play a dominant role in particulate fouling. In addition, it became clear that the predictive value of MFI measured at constant pressure was limited. For these reasons, MFI test measured at constant flux (with ultra-filtration membranes) was eventually developed over the last decade (Boerlage et al., 2004; Salinas Rodriguez et al., 2012). A comparison in the reduction of fouling potential as measured by MFI-UF in UF and GMF pretreatment systems is presented in Table 2-5. In general, UF membranes are superior over GMF in terms of MFI-UF reduction.

Table 2-5 Reported reduction in particulate/colloidal fouling potential as measured by MFI-UF for various pretreatment processes (Source: Salinas Rodriguez, 2011)

Treatment	Source water	Remarks	Reduction (%)
Granular media filtration	W. Mediterranean Sea	Anthracite-sand; 1.5 mg Fe/L	19
	N. Mediterranean Sea	Anthracite-sand; 2 mg Fe/L	37
Ultrafiltration	W. Mediterranean Sea	PVDF, 0.03 μm	66
	W. Mediterranean Sea	PVDF, 0.02 μm	52
	N. Mediterranean Sea	PVDF; 0.01 μm	68
	North Sea	PES 300 kDa 0.5 mg Al/L	88

2.4.5 Biological fouling potential

Measuring biological fouling potential of RO feed water is rather complicated. Over the years, multiple parameters have been proposed as indicators of biofouling potential, namely: adenosinetriphosphate (ATP), assimilable organic carbon (AOC) and biodegradable dissolved organic carbon (BDOC) (Vrouwenvelder and van der Kooij, 2001; Amy et al., 2011). So far, these parameters are mainly applied in non-saline waters and still not extensively used in seawater RO plants. Furthermore, inline monitors such as the biofilm monitor and membrane fouling simulator (MFS) have been introduced to measure biofilm formation rate (Vrouwenvelder and van der Kooij, 2001; Vrouwenvelder et al., 2006). Meanwhile, Liberman and Berman (2006) proposed a set of tests to determine the microbial support capacity (MSC) of water samples, namely chlorophyll-a, TEP, bacterial activity, total bacterial count, inverted microscope observations of sedimented water samples, biological oxygen demand (BOD), total phosphorous and total nitrogen. More investigations are needed to assess the reliability of these parameters/monitors to predict the biofouling potential of algal bloom impaired seawater.

2.5 Proposed strategies to control algae and AOM fouling

Reverse osmosis plants operating with direct/open source intake require extensive pretreatment of the raw water to maintain or prolong reliable performance and membrane life. To mitigate the adverse effects of algal blooms, a reliable pretreatment should continuously produce high quality RO feed water while maintaining stable operation. For example in GMF and UF pretreatment systems, a stable operation is based on the ability of the system to maintain acceptable backwash frequency at minimum chemical and energy requirement.

Following the 2008-2009 red tide outbreak in the Gulf of Oman which led to the shutdown of several desalination plants in the region, an expert workshop was held in Oman in 2012 on the impact of red tides and HABs on desalination operation. During this workshop, DAF and UF were highly recommended as possible alternatives to GMF for maintaining reliable operation in RO plants during severe algal bloom situations (Anderson and McCarthy, 2012). Installing a sub-surface intake (e.g., beach wells) instead of an open intake has been recently considered as a pretreatment option for SWRO. Figure 2-7 illustrates the different proposed process schemes based on existing technology as measures for mitigating the adverse effects of algal blooms in SWRO desalination systems.

Raw water abstraction in SWRO plants can be through an open or a sub-surface intake structure. For areas prone to algal blooms, sub-surface intake such as beach wells is preferred as it can serve as a natural slow sand filter which can allow substantial removal of algae and algal organic matter (Missimer et al., 2013). Consequently, less-extensive pretreatment processes are needed to maintain stable operation in the SWRO plant. However, sub-surface intakes are not applicable in some coastal locations where the geology of the area makes it unfeasible to install such intakes.

For SWRO plants operating with an open intake, primary and secondary pretreatment systems are installed to ensure acceptable RO feed water quality and stable operation during algal blooms. Primary pretreatment typically includes microstraining/screening to remove large suspended materials (> 50 μm), followed by coagulation and sedimentation or DAF. Secondary pretreatment typically comprises granular (dual) media filtration and/or UF. Granular media filtration requires a full coagulation/flocculation step prior to filtration.

Figure 2-7 Pretreatment options and process schemes to minimise fouling in SWRO system during algal blooms

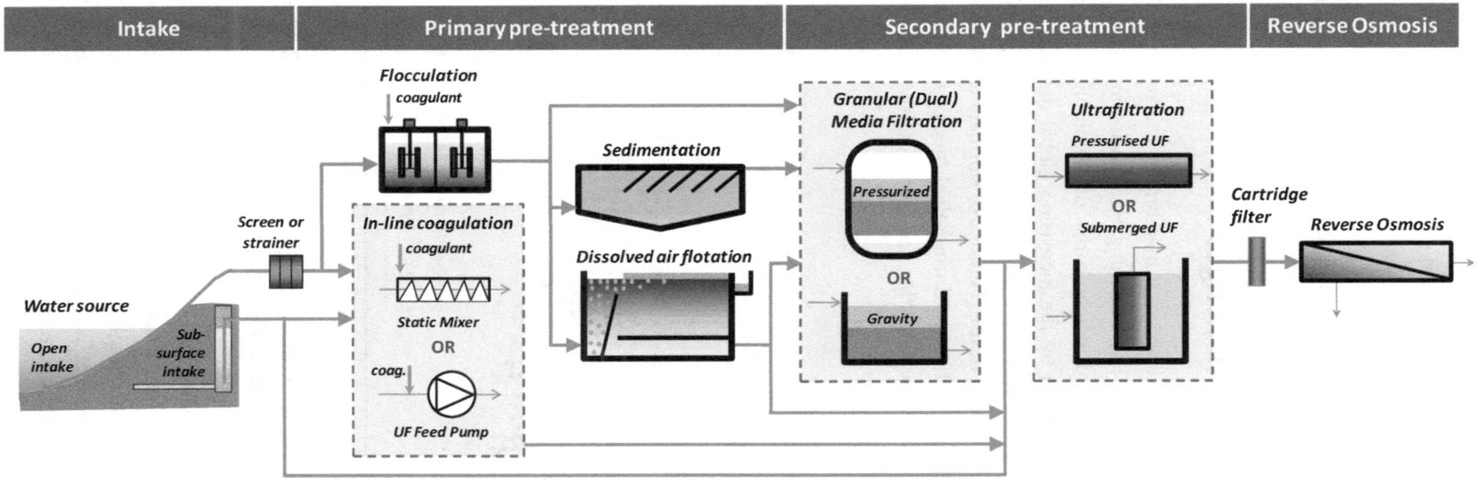

Process Options

Scheme A: Open intake → microstraining → coag-flocculation → sedimentation → granular media filtration → ultrafiltration → cartridge filtration → RO

Scheme B: Open intake → microstraining → coag-flocculation → sedimentation → granular media filtration → ultrafiltration → cartridge filtration → RO

Scheme C: Open intake → microstraining → coag-flocculation → dissolved air flotation → granular media filtration → ultrafiltration → cartridge filtration → RO

Scheme D: Open intake → microstraining → coag-flocculation → dissolved air flotation → ultrafiltration → cartridge filtration → RO

Scheme E: Open intake → microstraining → inline coagulation → ultrafiltration → cartridge filtration → RO

Scheme F: Subsurface intake → inline coagulation → ultrafiltration → cartridge filtration → RO

Scheme G: Subsurface intake → ultrafiltration → cartridge filtration → RO

On the other hand, UF can operate with or without coagulation and a flocculation basin or floc removal step is not necessarily required during coagulation. Schurer et al. (2013) demonstrated that UF is capable of maintaining stable operation during algal blooms when preceded by in-line coagulation without additional primary pretreatment. Other operational measures such as decreasing membrane flux and applying a forward flush cleaning may also improve the performance of UF during severe algal bloom situations.

2.6 Future pretreatment challenges

Membrane fouling and associated chemical cleaning of membranes is a major challenge to the cost-effective application of this technology. High cleaning frequencies make RO systems less reliable and robust, i.e., longer downtime and increased risk of membrane damage. This is in particular a concern for large plants. Driven by the increasing global demand as well as the economy of scale on the cost of desalinated water, it is projected that more large-scale RO plants (> 500,000 m^3/day) will be installed in the near future (Figure 2-8; Kurihara & Hanakawa, 2013). If pretreatment systems in these plants are ineffective during algal bloom, it can likely result in severe organic/biological fouling in SWRO which often requires extensive chemical cleanings (i.e., CIP) to restore membrane permeability.

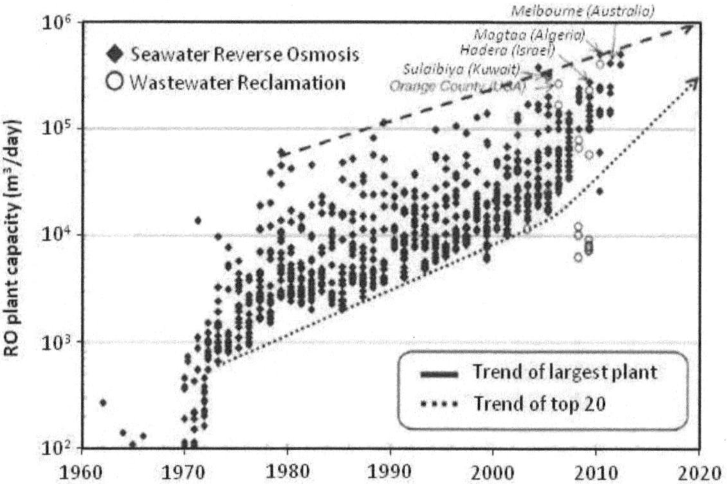

Figure 2-8 Productivity of the top 20 largest RO plants in the world from 1960-2012 and projected plant size for year 2020 (Kurihara and Hanakawa, 2013)

A CIP frequency of once per year is frequently considered as being convenient for large to extra-large SWRO plants to minimize costs, downtime and membrane damage. If pretreatment is not adequate high cleaning frequency will lead to high operational costs and ultimately membrane replacement will pose a large cost component for such plants. Moreover, operators may be charged with penalties if they fail to supply the contracted amounts of water. Consequently, for current and future large and extra-large SWRO plants, it seems to be essential that pretreatment systems are reliable in maintaining RO feed water with very low fouling potential. The increasing number of SWRO plants equipped with UF as pretreatment might reflect this expectation.

To minimise the cleaning frequency of SWRO plants affected by algal blooms, the development of new generation of pretreatment technology should focus on complete removal of algae and their AOM as well as limiting the concentration of nutrients in RO feed water. Removal of algae and AOM (e.g., TEP) in itself can eliminate organic fouling and may substantially delay the onset of biological fouling in SWRO as there is no "conditioning layer" to jump-start biofilm development. On the other hand, removal of essential nutrients (e.g., phosphate, AOC) from the RO feed water can further control biological growth in the system. Combining the two treatment strategies can potentially eliminate organic/biological fouling in seawater RO systems, even in algal bloom situations.

2.7 Summary and outlook

The recent severe algal bloom outbreaks in the Middle East have resulted in temporary closure of various seawater desalination installations in the region, mainly due to breakdown of pretreatment systems and/or as a drastic measure to prevent irreversible fouling problems in downstream RO membranes. The major issues which may occur in SWRO plants during an algal bloom are: (1) particulate/organic fouling of pretreatment systems (e.g., GMF, MF/UF) by algal cells, their detritus and/or AOM, and (2) biological fouling of NF/RO initiated and/or enhanced by AOM. As membrane-based desalination is expected to grow in terms of production capacity and global application, many coastal areas in the world will potentially face a similar scenario in the future if these problems are not addressed adequately.

To tackle serious operational problems of membrane-based desalination plants, several pretreatment strategies are currently being proposed. SWRO plants that are fitted with conventional granular media filtration, can gain significant benefits in terms of capacity and product water quality, if preceded by DAF. As a consequence, coagulant consumption will increase as a result of coagulant dosage prior to DAF. Therefore, particularly in countries with stringent legislation, such schemes should be foreseen with sludge handling and/or treatment facilities. If conditions allow, UF pretreatment should be incorporated in SWRO plant design as pretreatment. The advantages of UF over conventional pretreatment systems are well known to the membrane and desalination society. Significant additional benefits of UF pretreatment systems can be gained during periods of severe algal bloom in terms of maintaining pretreatment capacity and RO feed water quality. Moreover, coagulant consumption is significantly lower in these systems as compared to conventional pretreatment systems. The future of UF pretreatment of SWRO lies in the development of tight UF membranes (low MWCO) that can deliver high quality RO feed water (very low SDI or MFI) at minimal coagulant consumption and high output rates.

2.8 Acknowledgements

This chapter was written in collaboration with Loreen Villacorte.

2.9 References

Alhadidi, A., Blankert, B., Kemperman, A.J.B., Schurer, R., Schippers, J.C., Wessling, M., van der Meer, W.G.J., 2013. Limitations, improvements and alternatives of the silt density index. Desalination and Water Treatment 51(4-6), 1104-1113.

Alldredge, A.L., Passow, U., Logan, B.E., 1993. The abundance and significance of a class of large, transparent organic particles in the ocean. Deep-Sea Research I 40, 1131–1140.

Amy, G.L., Salinas-Rodriguez, S.G., Kennedy, M.D., Schippers, J.C., Rapenne, S., Remize, P.J., Barbe, C., Manes, C.L.D.O., West, N.J., Lebaron, P., van der Kooij, D., Veenendaal, H., Schaule, G., Petrowski, K., Huber, S., Sim, L.N., Ye, Y., Chen, V., Fane, A.G., 2011. Water quality assessment tools. In: Drioli, E., Criscuoli, A., Macedonio, F. (Eds.) Membrane-Based Desalination - An Integrated Approach (MEDINA). IWA Publishing, New York, 3-32.

Anderson, D.M., McCarthy, S., 2012. Red tides and harmful algal blooms: Impacts on desalination operations. Middle East Desalination Research Center, Muscat, Oman. Online: www.medrc.org/download/habs_and_desaliantion_workshop_report_final.pdf

Arruda-Fatibello, S.H.S., Henriques-Vieira, A.A., Fatibello-Filho, O., 2004. A rapid spectrophotometric method for the determination of transparent exopolymer particles (TEP) in freshwater. Talanta 62(1), 81-85.

Azetsu-Scott, K., Passow, U., 2004. Ascending marine particles: Significance of transparent exopolymer particles (TEP) in the upper ocean. Limnol. Oceanogr 49(3), 741-748.

Baker, J.S., Dudley, L.Y., 1998. Biofouling in membrane systems - a review. Desalination 118(1-3), 81-89.

Bar-Zeev, E., Berman-Frank, I., Girshevitz, O., Berman, T., 2012a. Revised paradigm of aquatic biofilm formation facilitated by microgel transparent exopolymer particles. PNAS 109(23), 9119-9124.

Bar-Zeev, E., Belkin, N., Liberman, B., Berman, T., Berman-Frank, I., 2012b. Rapid sand filtration pretreatment for SWRO: Microbial maturation dynamics and filtration efficiency of organic matter. Desalination 286, 120-130.

Bar-Zeev, E., Belkin, N., Liberman, B., Berman-Frank, I., Berman, T., 2013. Bioflocculation: Chemical free, pretreatment technology for the desalination industry. Water Research 47(9), 3093-3102.

Bar-Zeev, E., Berman-Frank, I., Liberman, B., Rahav, E., Passow, U., Berman, T., 2009. Transparent exopolymer particles: Potential agents for organic fouling and biofilm formation in desalination and water treatment plants. Desalination and Water Treatment 3(1-3), 136-142.

Berktay, A., 2011. Environmental approach and influence of red tide to desalination process in the middle-east region. International Journal of Chemical and Environmental Engineering 2(3), 183-188.

Berman, T., Holenberg, M., 2005. Don't fall foul of biofilm through high TEP levels. Filtration & Separation 42(4), 30-32.

Berman, T., 2010. Biofouling: TEP - a major challenge for water filtration. Filtration & Separation 47(2), 20-22.

Berman, T., Mizrahi, R., Dosoretz, C.G., 2011. Transparent exopolymer particles (TEP): A critical factor in aquatic biofilm initiation and fouling on filtration membranes. Desalination 276(1-3), 184-190.

Boerlage, S.F.E., Kennedy, M.D., Aniye, M.P., Abogrean, E.M., El-Hodali, D.E.Y., Tarawneh, Z.S., Schippers, J.C., 2000. Modified Fouling Index - ultrafiltration to compare pretreatment processes of reverse osmosis feed water. Desalination 131(1-3), 201-214.

Boerlage, S.F.E., Kennedy, M., Tarawneh, Z., De Faber, R., Schippers, J.C., 2004. Development of the MFI-UF in constant flux filtration. Desalination 161(2), 103-113.

Brehant, A., Bonnelye, V., Perez, M., 2002. Comparison of MF/UF pretreatment with conventional filtration prior to RO membranes for surface seawater desalination. Desalination 144, 353-360.

Caron , D.A., Garneau, M.E., Seubert, E., Howard M.D.A., Darjany L., Schnetzer A., Cetinic, I., Filteau, G., Lauri, P., Jones, B., Trussell, S., 2010. Harmful Algae and Their Potential Impacts on Desalination Operations of Southern California. Water Research 44, 385–416.

Castaing, J.B., Masse, A., Sechet, V., Sabiri, N.E., Pontie, M., Haure, J., Jaouen, P., 2011. Immersed hollow fibres microfiltration (MF) for removing undesirable micro-algae and protecting semi-closed aquaculture basins. Desalination 276(1-3), 386-396.

Chin, W.C., Orellana, M.V., Verdugo, P., 1998. Spontaneous assembly of marine dissolved organic matter into polymer gels. Nature 391, 568–572.

DesalData (2013) Worldwide desalination inventory (MS Excel Format), downloaded from DesalData.com (GWI/IDA) on June 6, 2013.

Dramas, L., Croué, J.P., 2013. Ceramic membrane as a pretreatment for reverse osmosis: interaction between marine organic matter and metal oxides. Desalination and Water Treatment 51(7-9), 1781-1789.

Edzwald, J.K., 2010. Dissolved air flotation and me. Water Research 44, 2077-2106.

Eshel, G., Elifantz, H., Nuriel, S., Holenberg, M., Berman, T., 2013. Microfiber filtration of lake water: impacts on TEP removal and biofouling development. Desalination and Water Treatment 51 (4-6), 1043-1049.

Flemming, H.C., 2002. Biofouling in water systems–cases, causes and countermeasures. Applied Microbiology and Biotechnology 59, 629-640.

Flemming, H.C., Schaule, G., Griebe, T., Schmitt, J., Tamachkiarowa, A., 1997. Biofouling-the Achilles heel of membrane processes. Desalination 113(2), 215-225.

Gamal Khedr, M., 2000. Membrane fouling problems in reverse osmosis desalination applications. Desalination and Water Reuse 10, 8-17.

Gille, D., Czolkoss, W., 2005. Ultrafiltration with multi-bore membranes as seawater pretreatment. Desalination 182(1-3), 301-307.

Glueckstern, P., Priel, M., Wilf, M., 2002. Field evaluation of capillary UF technology as a pretreatment for large seawater RO systems. Desalination 147, 55-62.

Gregory R., Edzwald J.K., 2010. Sedimentation and flotation, J.K. Edzwald (Ed.), Water Quality and Treatment (Sixth Ed.), McGraw Hill, New York (2010) (Chapter 9).

Gregory, R., 1997. Summary of General Developments in DAF for Water Treatment since 1976. Proceedings Dissolved Air Flotation Conference. The Chartered Institution of Water and Environmental Management, London, 1-8.

Guastalli, A. R., Simon, F. X., Penru, Y., de Kerchove, A., Llorens, J., and Baig, S. (2013). Comparison of DMF and UF pretreatments for particulate material and dissolved organic matter removal in SWRO desalination. Desalination 322, 144-150.

Hallé, C., Huck, P.M., Peldszus, S., Haberkamp, J., Jekel, M., 2009. Assessing the performance of biological filtration as pretreatment to low pressure membranes for drinking water. Environmental Science and Technology 43(10), 3878-3884.

Halpern, D.F., McArdle, J., Antrim, B., 2005. UF pretreatment for SWRO: pilot studies. Desalination 182, 323-332.

Heijman, S.G.J., Kennedy, M.D., van Hek, G.J., 2005. Heterogeneous fouling in dead-end ultrafiltration. Desalination 178(1-3), 295-301.

Heijman, S.G.J., Vantieghem, M., Raktoe, S., Verberk, J.Q.J.C., van Dijk, J.C., 2007. Blocking of capillaries as fouling mechanism for dead-end ultrafiltration. Journal of Membrane Science 287(1), 119-125.

Henderson, R.K., Parsons, S.A., Jefferson, B., 2009. The potential for using bubble modification chemicals in dissolved air flotation for algae removal. Separation Science and Technology 44(9), 1923-1940.

Henderson, R., Parsons, S.A., Jefferson, B., 2008. The impact of algal properties and pre-oxidation on solid-liquid separation of algae. Water Research 42(8-9), 1827-1845.

Huang, G., Meng, F., Zheng, X., Wang, Y., Wang, Z., Liu, H., Jekel, M., 2011. Biodegradation behavior of natural organic matter (NOM) in a biological aerated filter (BAF) as a pretreatment for ultrafiltration (UF) of river water. Applied Microbiology and Biotechnology 90(5), 1795-1803.

Huber, S.A., Balz, A., Abert, M., Pronk, W., 2011. Characterisation of aquatic humic and non-humic matter with size-exclusion chromatography - organic carbon detection - organic nitrogen detection (LC-OCD-OND). Water Research 45, 879-885.

Kennedy, M.D., Muñoz Tobar, F.P., Amy, G., Schippers, J.C., 2009. Transparent exopolymer particle (TEP) fouling of ultrafiltration membrane systems. Desalination and Water Treatment 6(1-3), 169-176.

Kim, S.H., Yoon, J.S., 2005. Optimization of microfiltration for seawater suffering from red-tide contamination. Desalination 182(1–3), 315–321.

Komlenic, R., Berman, T., Brant, J.A., Dorr, B., El-Azizi, I., Mowers, H., 2013. Removal of polysaccharide foulants from reverse osmosis feed water using electroadsorptive cartridge filters. Desalination and Water Treatment 51(4-6), 1050-1056.

Kurihara, M., Hanakawa, M., 2013. Mega-ton Water System: Japanese national research and development project on seawater desalination and wastewater reclamation. Desalination 308(0), 131-137.

Ladner, D.A., Vardon, D.R., Clark M.M., 2010. Effects of Shear on Microfiltration and Ultrafiltration Fouling by Marine Bloom-forming Algae. Journal of Membrane Science 356, 33-43.

Laycock, M.V., Anderson, D.M., Naar, J., Goodman, A., Easy, D.J., Donovan, M.A., Li, A., Quilliam, M.A., Al Jamali, E., Alshihi, R., Alshihi, R., 2012. Laboratory desalination experiments with some algal toxins. Desalination 293, 1-6.

Le Gallou, S., Bertrand, S., Madan, K.H., 2011. Full coagulation and dissolved air flotation: a SWRO key pretreatment step for heavy fouling seawater. In: Proceedings of International Desalination Association World Congress, Perth, Australia.

Lerch, A., Uhl, W., Gimbel, R., 2007. CFD modelling of floc transport and coating layer build-up in single UF/MF membrane capillaries driven in inside-out mode. Water Science and Technology: Water Supply 7(4), 37-47.

Liberman, B., Berman, T., 2006. Analysis and monitoring: MSC - a biologically oriented approach. Filtration & Separation 43(4), 39-40.

Longhurst, S.J., Graham, N.J.D., 1987. Dissolved air flotation for potable water treatment: a survey of operational units in Great Britain. The Public Health Engineer 14(6), 71-76.

Missimer, T.M., Ghaffour, N., Dehwah, H.A., Rachman, R., Maliva, R.G., Amy, G., 2013. Subsurface intakes for seawater reverse osmosis facilities: Capacity limitation, water quality improvement, and economics. Desalination 322, 37–51.

Myklestad, S.M., 1995. Release of extracellular products by phytoplankton with special emphasis on polysaccharides. Science of the Total Environment 165, 155-164.

Naidu, G., Jeong, S., Vigneswaran, S., Rice, S.A., 2013. Microbial activity in biofilter used as a pretreatment for seawater desalination. Desalination 309, 254-260.

Nazzal, N., 2009. 'Red tide' shuts desalination plant. Gulf News, Dubai, UAE. Available from http://gulfnews.com/news/gulf/uae/environment/red-tide-shuts-desalination-plant-1.59095.

Panglisch, S., 2003. Formation and prevention of hardly removable particle layers in inside-out capillary membranes operating in dead-end mode. Water Science and Technology: Water Supply 3(5-6), 117-124.

Pankratz, T., 2008. Red Tides Close Desal Plants, Water Desalination Report, 44 (44).

Park, K.S., Mitra, S.S., Yim, W.K., Lim, S.W., 2013. Algal bloom - critical to designing SWRO pretreatment and pretreatment as built in Shuwaikh, Kuwait SWRO by Doosan. Desalination and Water Treatment 51(31-33), 1-12.

Passow, U., Alldredge, A.L., 1995. A dye-binding assay for the spectrophotometric measurement of transparent exopolymer particles (TEP). Limnol. Oceanogr 40(7), 1326-1335.

Passow, U., 2000. Formation of transparent exopolymer particles (TEP) from dissolved precursor material. Marine Ecology Progress Series 192, 1-11.

Passow, U., Alldredge, A.L., 1994. Distribution, size, and bacterial colonization of transparent exopolymer particles (TEP) in the ocean. Marine Ecology Progress Series 113, 185–198.

Pearce, G., 2009. SWRO pretreatment: treated water quality. Filtration & Separation 46(6), 30-33.

Pearce, G.K., 2007. The case for UF/MF pretreatment to RO in seawater applications. Desalination 203(1-3), 286-295.

Peperzak, L., Poelman, M., 2008. Mass mussel mortality in the Netherlands after a bloom of phaeocystis globosa (prymnesiophyceae). Journal of Sea Research 60(3), 220-222.

Plantier, S., Castaing, J.B., Sabiri, N.E., Massé, A., Jaouen, P., Pontié, M., 2012. Performance of a sand filter in removal of algal bloom for SWRO pretreatment. Desalination and Water Treatment 51(7-9), 1838-1846.

Qu, F., Liang, H., He, J., Ma, J., Wang, Z., Yu, H., Li, G., 2012a. Characterization of dissolved extracellular organic matter (dEOM) and bound extracellular organic matter (bEOM) of microcystis aeruginosa and their impacts on UF membrane fouling. Water Research 46(9), 2881-2890.

Qu, F., Liang, H., Tian, J., Yu, H., Chen, Z., Li, G., 2012b. Ultrafiltration (UF) membrane fouling caused by cyanobateria, fouling effects of cells and extracellular organics matter (EOM). Desalination 293, 30-37.

Richlen, M.L., Morton, S.L., Jamali, E.A., Rajan, A., Anderson, D.M., 2010. The Catastrophic 2008-2009 red tide in the Arabian Gulf region, with observations on the identification and phylogeny of the fish-killing dinoflagellate cochlodinium polykrikoides. Harmful Algae 9(2), 163-172.

Rovel, J.M., 2003. Why a SWRO in Taweelah - pilot plant results demonstrating feasibility and performance of SWRO on Gulf water? In: Proceedings of International Desalination Association World Congress, Nassau, Bahamas.

Salinas Rodríguez, S.G., Kennedy, M.D., Schippers, J.C., Amy, G.L., 2009. Organic foulants in estuarine and bay sources for seawater reverse osmosis - Comparing pretreatment processes with respect to foulant reduction. Desalination and Water Treatment 9 (1-3), 155-164.

Salinas-Rodriguez, S.G., 2011. Particulate and organic matter fouling of SWRO systems: Characterization, modelling and applications. Doctoral dissertation, UNESCO-IHE/TUDelft, Delft.

Sanz, M.A., Guevara, D., Beltrán, F., Trauman, E., 2005. 4 Stages pretreatment reverse osmosis for South-Pacific seawater: El Coloso plant (Chile). In: Proceedings of International Desalination Association World Congress, Singapore.

Schippers, J.C., Verdouw, J., 1980. The modified fouling index, a method of determining the fouling characteristics of water. Desalination 32, 137-148.

Schurer, R., Tabatabai, A., Villacorte, L., Schippers, J.C., Kennedy, M.D., 2013. Three years operational experience with ultrafiltration as SWRO pretreatment during algal bloom. Desalination and Water Treatment 51 (4-6), 1034-1042.

Schurer, R., Janssen, A., Villacorte, L., Kennedy, M.D., 2012. Performance of ultrafiltration and coagulation in an UF-RO seawater desalination demonstration plant. Desalination and Water Treatment 42(1-3), 57-64.

Teixeira, M.R., Rosa, M.J., 2007. Comparing dissolved air flotation and conventional sedimentation to remove cyanobacterial cells of microcystis aeruginosa. Part II. The effect of water background organics. Separation and Purification Technology 53(1), 126-134.

Teixeira, M.R., Sousa, V., Rosa, M.J., 2010. Investigating dissolved air flotation performance with cyanobacterial cells and filaments. Water Research 44(11), 3337-3344.

Thornton, D.C.O., Fejes, E.M., DiMarco, S.F., Clancy, K.M., 2007. Measurement of acid polysaccharides (APS) in marine and freshwater samples using alcian blue. Limnol. Oceanogr 5, 73-87.

van Nevel, S., Hennebel, T., De Beuf, K., Du Laing, G., Verstraete, W., Boon, N., 2012. Transparent exopolymer particle removal in different drinking water production centers. Water Research 46(11), 3603-3611.

van Puffelen, J., Buijs, P.J., Nuhn, P.N.A.M., Hijen, W.A.M., 1995. Dissolved air flotation in potable water treatment: the Dutch experience. Water Science and Technology, 31(3-4), 149-157.

Verdugo, P., Alldredge, A.L., Azam, F., Kirchman, D.L., Passow, U., Santschi, P.H., 2004. The oceanic gel phase: a bridge in the DOM–POM continuum. Marine Chemistry 92, 67-85.

Villacorte, L.O., Kennedy, M.D., Amy, G.L., Schippers, J.C., 2009a. The fate of transparent exopolymer particles (TEP) in integrated membrane systems: removal through pretreatment processes and deposition on reverse osmosis membranes. Water Research 43(20), 5039-5052.

Villacorte, L.O., Kennedy, M.D., Amy, G.L., Schippers, J.C., 2009b. Measuring transparent exopolymer particles (TEP) as indicator of the (bio)fouling potential of RO feed water. Desalination and Water Treatment 5, 207-212.

Villacorte, L.O., Schurer, R., Kennedy, M., Amy, G., Schippers, J.C., 2010a. The fate of transparent exopolymer particles in integrated membrane systems: a pilot plant study in Zeeland, The Netherlands. Desalination and Water Treatment 13, 109-119.

Villacorte, L.O., Schurer, R., Kennedy, M., Amy, G., Schippers, J.C., 2010b. Removal and deposition of transparent exopolymer particles (TEP) in seawater UF-RO system. IDA Journal 2 (1), 45-55.

Villacorte, L.O., 2014. Algal blooms and membrane based desalination technology. Doctoral dissertation, UNESCO-IHE/TUDelft, Delft.

Villacorte, L.O., Ekowati, Y., Winters, H., Amy, G.L., Schippers, J.C., Kennedy, M.D., 2013. Characterisation of transparent exopolymer particles (TEP) produced during algal bloom: a membrane treatment perspective. Desalination and Water Treatment 51 (4-6), 1021-1033.

Vlaski A., 1997. Microcystis aeruginosa Removal by Dissolved Air Flotation (DAF): Options for Enhanced Process Operation and Kinetic Modelling. Doctoral dissertation, IHE/TUD, Delft.

Vrouwenvelder, J.S., van der Kooij, D., 2001. Diagnosis, prediction and prevention of biofouling of NF and RO membranes. Desalination 139(1-3), 65-71.

Vrouwenvelder, J.S., van Paassen, J.A.M., Wessels, L.P., van Dam, A.F., Bakker, S.M., 2006. The membrane fouling simulator: A practical tool for fouling prediction and control. Journal of Membrane Science 281(1-2), 316-324.

WaterWorld, 2013. Fujairah hybrid desalination plant to expand with dissolved air flotation system. Available at: http://www.waterworld.com/articles/2013/01/fujairah-hybrid-desalination-plant-to-expand-with-dissolved-air-floatation-system.html.

WHO, 2007. Desalination for Safe Water Supply, Guidance for the Health and Environmental Aspects Applicable to Desalination, World Health Organization (WHO), Geneva.

Wilf, M., Schierach, M.K., 2001. Improved performance and cost reduction of RO seawater systems using UF pretreatment. Desalination 135, 61-68.

Winters, H., Isquith, I.R., 1979. In-plant microfouling in desalination. Desalination 30(1), 387-399.

Wolf, P.H., Siverns, S., Monti, S., 2005. UF membranes for RO desalination pretreatment. Desalination 182, 293-300.

Zhang, Y., Tian, J., Nan, J., Gao, S., Liang, H., Wang, M., Li, G., 2011. Effect of PAC addition on immersed ultrafiltration for the treatment of algal-rich water. Journal of Hazardous Materials 186(2-3), 1415-1424.

3

FOULING POTENTIAL OF COAGULANT IN MF/UF SYSTEMS

This chapter investigates the extent to which aluminium as coagulant contributes to fouling in microfiltration membranes in one filtration cycle. Using a model based on the Carman-Kozeny equation, the effect of particle size, density and cake/gel layer porosity on fouling potential and the resulting pressure increase in microfiltration membranes in one filtration cycle was calculated. Experiments were carried out to determine the fouling potential - as measured by the Modified Fouling Index (MFI) at constant flux - of polyaluminium chloride aggregates/flocs formed under various conditions of dose, pH, mixing and flocculation. Theoretical calculations indicated that the contribution of primary particles larger than a few nanometres to the rate of fouling of a microfiltration membrane, for flux values of up to 150 L/m^{2}h, was marginal. Prediction of pressure development in one filtration cycle in microfiltration membranes based on experimentally obtained MFI values showed that polyaluminium chloride aggregates formed under given coagulation conditions did not contribute significantly to pressure increase. The fouling potential of aluminium aggregates varied with varying coagulation conditions; higher MFI was observed at lower coagulant dose particularly at pH values higher than 6.

This chapter is based on:

Tabatabai, S.A.A., Kennedy, M.D., Amy, G.L., Schippers, J.C., 2009. Optimizing inline coagulation to reduce chemical consumption in MF/UF systems. Desalination and Water Treatment 6, 94-101.

3.1 Introduction

Coagulation is commonly applied in conventional water treatment processes such as sedimentation/flotation followed by media filtration to improve process performance, e.g., in terms of turbidity removal and surface loading rates. Considerable efforts have been made in over half a century of research and implementation to establish theoretical understanding of the coagulation process and to define optimum design criteria. Process conditions that affect coagulation are dose, mixing intensity (G), flocculation Gt, pH and temperature. Velocity gradient (G) and Gt are essential parameters in the design and operation of coagulation systems (Binnie et al., 2002). Camp and Stein (1943) developed the basic theory of power input for mixing and defined G as the mean velocity gradient which is proportional to the square root of power dissipated per unit volume of liquid.

Upon coagulant addition, hydrolysis is instantaneous and the reactions that lead to the formation of reactive coagulant species occur immediately, resulting in subsequent polymerization and precipitation of metal hydroxide. Species that are required for charge neutralization reactions to occur are reportedly formed in less than 0.1 s (van Benschoten and Edzwald, 1990) if no hydrolysis polymers are formed and within 1 second if polymers are formed (Hahn and Stumm, 1968), while formation of aluminium hydroxide precipitate may occur between 1-7 s (Letterman et al., 1973; Amirtharajah and Mills, 1982). Consequently, mixing has to ensure that the coagulant is fully dispersed within the liquid in the shortest time possible. In literature, G-values in the range of 250 - 10,000 s^{-1} are reported. When mixing is applied in a tank, G-values are typically in the lower range and mixing time is in the order of minutes. For inline mixers, the higher range is encountered and mixing times may be as short as a few seconds (Binnie et al., 2002). USEPA recommends a mixing time of at least one minute and G-values of 300 - 1000 s^{-1} (EPA, 1999).

Flocculation occurs by particle collision through thermally induced Brownian motion (perikinetic aggregation), stirring or differential settling (orthokinetic aggregation). For orthokinetic aggregation, Camp (1955) recommended G-values in the range of 20 - 75 s^{-1} and a Gt range of 20,000 - 200,000 for optimum flocculation when large flocs are required prior to clarification, e.g., sedimentation. Flocculators are usually designed for G-values in the range of 10 to 60 s^{-1} and Gt values in the range of 10,000 to 100,000 for conventional water treatment schemes (Bratby, 2006).

Advanced treatment options such as micro- and ultrafiltration (MF/UF) for municipal water treatment and as pretreatment to reverse osmosis (RO) are not reliant on coagulation for turbidity and particle removal (for disinfection). However many MF/UF systems treating surface water can only be operated at very low permeate fluxes to ensure low fouling rates. While operation at low filtration flux can control fouling in MF/UF systems, large investment in membrane surface area is required to meet production capacity. Under such conditions, coagulation is applied to control fouling in MF/UF membranes while maintaining relatively high permeate flux values. MF/UF performance is generally characterized in terms of hydraulic performance and permeate quality. Four main aspects of the hydraulic performance (Figure 3-1a) of MF/UF systems are identified as,

- Pressure development in subsequent filtration cycles;
- Permeability loss after backwashing, i.e., non-backwashable fouling;
- Permeability loss after chemically enhanced backwashing cleaning (CEB), i.e., irreversible fouling (I);
- Permeability loss after cleaning in place (CIP), i.e., irreversible fouling (II).

It must be noted that at high fouling rates and compressibility, filtration curves for successive filtration cycles deviate from linear. Coagulation improves MF/UF process performance by,

- Reducing total pressure increase in each filtration cycle by modifying particle and cake/gel layer properties;
- Enhancing permeability recovery after backwashing by avoiding adsorption into and blocking of membrane pores (Guigui et al., 2002; Hwang and Liu, 2002). These effects are illustrated in Figure 3-1(b).

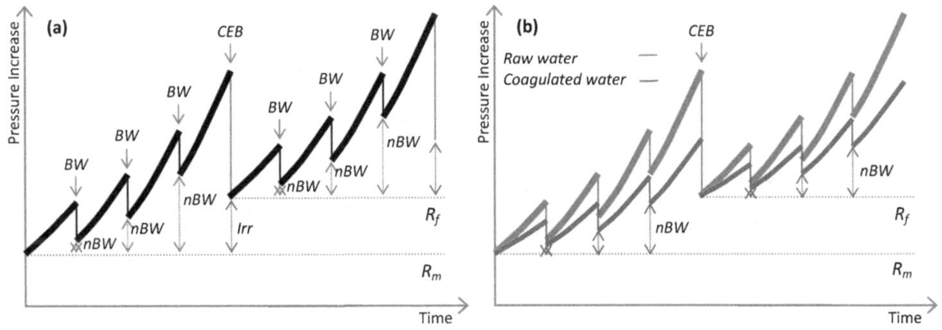

Figure 3-1 Schematic presentation of pressure development and fouling in dead-end MF/UF systems (a) raw feed water and (b) effect of coagulation. BW is backwash; nBW is non-backwashable fouling, Irr is irreversible fouling (I), R_m is clean membrane resistance and R_f is membrane resistance after fouling.

Several coagulation modes may be applied in coagulation prior to MF/UF (Figure 3-2),

a) Rapid mixing and no flocculation, i.e., inline coagulation
b) Rapid mixing and flocculation
c) Rapid mixing, flocculation and sedimentation
d) Rapid mixing, flocculation and flotation

This study focuses on modes (a) inline coagulation, and (b) rapid mixing and flocculation. Inline coagulation is primarily characterized by the absence of a floc removal step, e.g., sedimentation or flotation. In this mode, flocculation is achieved during the mixing step (simultaneously with coagulant dispersion and hydrolysis) in e.g., static or mechanical mixers, in the pump chamber of the MF/UF pump or in the piping network that carries the coagulated feed water to the membranes.

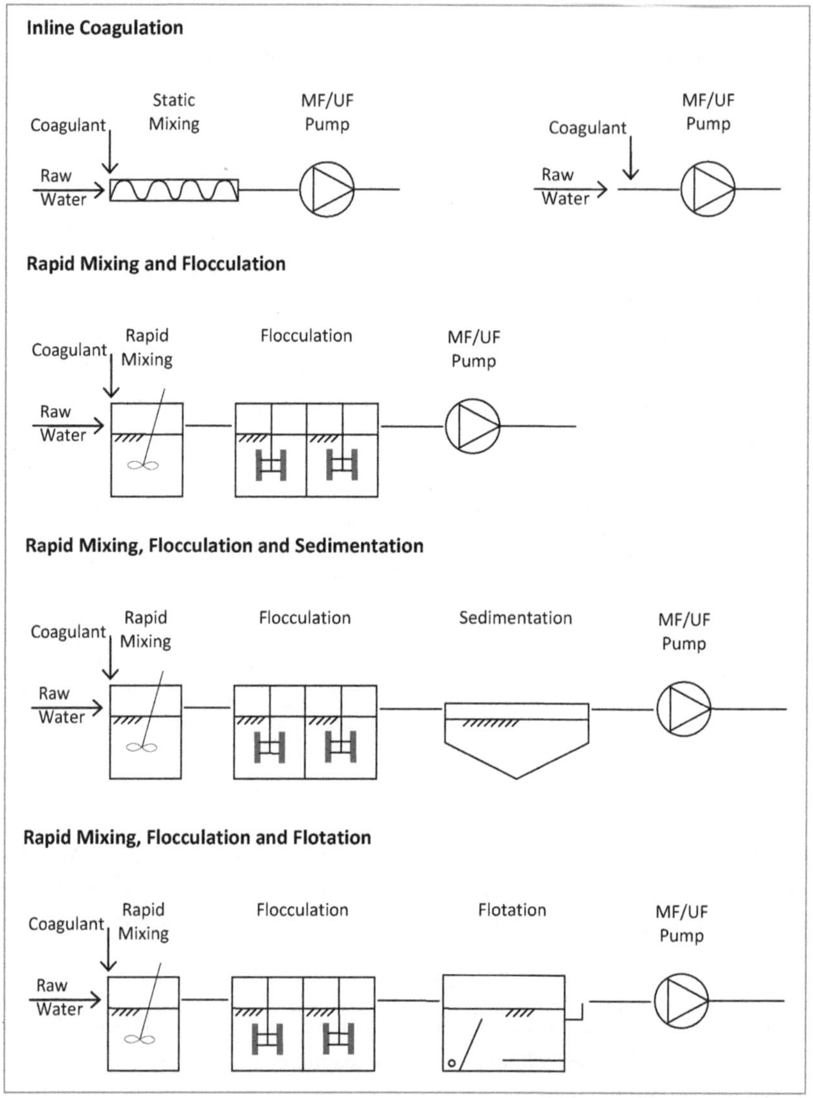

Figure 3-2 Schematic presentation of different modes of coagulant application prior to MF/UF systems

Although coagulation is successfully applied in MF/UF systems to control fouling, high coagulant doses are frequently encountered in such systems. The actual values of dose and contact time and the extent to which coagulation can improve MF/UF performance is not clear, as existing literature on coagulation (theory, optimum process conditions and assessment criteria) mostly addresses conventional processes.

Ferric and aluminium salts are commonly applied in MF/UF systems. Two forms of aluminium salts are used for coagulation, namely Al(III), e.g., aluminium sulphate, and partially hydrolyzed polymeric aluminium salts, e.g., poly aluminium chloride (PACl). The polymeric form of aluminium has a much higher rate of floc formation, which makes it attractive at lower temperatures and eliminates the need for coagulant aid. However aluminium applied in MF/UF systems as RO pretreatment poses a significant risk of RO membrane fouling (Gabelich et al., 2006; Ventresque et al., 2000). Ferric salts do not carry this risk, due to much lower solubility. However, in some cases, non-backwashable fouling of MF/UF membranes has been reported when ferric is applied, resulting in the need for chemical cleaning, e.g., at low pH or with oxalic/ascorbic acid (Schurer et al., 2012; 2013).

The beneficial effect of dosing coagulants in front of MF/UF has been demonstrated frequently. However the contribution of the coagulant itself on the remaining (expectedly reduced) fouling potential, reflected in slower pressure development in filtration cycles, is unknown. This study aimed at determining the contribution of PACl itself on fouling potential in MF/UF membranes by,

- Quantifying - theoretically - the influence of particle properties such as size, density, and cake/gel layer porosity on fouling potential of aggregates; and predicting pressure development in MF/UF membranes operated at constant flux.
- Determining the contribution of polyaluminium chloride (PACl) itself under different conditions of dose, pH, G and Gt to fouling potential in subsequent filtration cycles in MF/UF membranes.
- Predicting pressure development in MF/UF systems operating at constant flux, fed with water carrying aluminium hydroxide aggregates formed under different conditions.

For this purpose filtration experiments were conducted at constant pressure with microfiltration membranes (0.1 μm) fed with water containing aluminium hydroxide aggregates formed under different coagulation conditions. Modified Fouling Index (MFI) - calculated from the development of flux as a function of time and total filtered volume - was used as an index of the fouling potential of aluminium hydroxide aggregates.

3.2 Theoretical background

3.2.1 Modified Fouling Index (MFI)

The Modified Fouling Index (MFI) was introduced by Schippers and Verdouw (1980) to measure the fouling potential of a feed water containing particles, for filtration at constant pressure. This test is based on the development of cake/gel filtration, soon after filtration is started. Filtration flux through a membrane can be given by Darcy's Law for flow through a porous medium and the resistance in series model,

$$J = \frac{dV}{Adt} = \frac{\Delta P}{\eta(R_m + R_b + R_c)} \qquad \text{(Eq. 3-1)}$$

Where,
J is filtration flux (L/m^2h)
η is fluid viscosity (Pa.s)
ΔP is applied pressure (bar)
R_m is clean membrane resistance (1/m)
R_b is resistance due to pore blocking (1/m)
R_c is cake/gel layer resistance (1/m).

Resistance of the cake/gel layer formed during constant pressure filtration is proportional to the amount of particles deposited on the surface of the medium, provided retention is constant throughout filtration, and is defined by Ruth et al. (1933),

$$R_c = \frac{V}{A}.I \qquad \text{(Eq. 3-2)}$$

Where,
I is fouling index (m^{-2}) and is defined as the product of bulk particle concentration (C_b) and specific cake/gel resistance (α).

Combining Eq. 3-1 and Eq. 3-2 and assuming that pore blocking does not play a role after a short period from the start of filtration,

$$\frac{dt}{dV} = \frac{\eta R_m}{A\Delta P} + \frac{\eta I}{\Delta P A^2} V \qquad \text{(Eq. 3-3)}$$

Integration at constant pressure assuming I is time independent and porosity remains uniform throughout the depth of the cake/gel (i.e., no compression of the cake/gel), results in the well known filtration equation,

$$\frac{t}{V} = \frac{\eta R_m}{A\Delta P} + \frac{\eta I}{2\Delta P A^2} V \qquad \text{(Eq. 3-4)}$$

When t/V is plotted against V, a graph such as the one shown in Figure 3-3 may be obtained, depending on the occurrence of different filtration mechanisms. The minimum gradient of the curve (tan α) is used to calculate MFI, based on reference conditions of pressure, temperature and membrane surface area,

$$\mathrm{MFI} = \frac{\eta_{20°C}.\tan\alpha}{2\Delta P_0 A_0^2} \qquad \text{(Eq. 3-5)}$$

Where,
$\tan\alpha$ is the minimum gradient of the t/V vs. V plot (s/L²)
$\eta_{20°C}$ is reference viscosity at 20 °C (0.001 Pa.s)
ΔP_o is reference pressure (2 bars)
A_o is reference membrane surface area (13.8 $*10^{-4}$ m²).

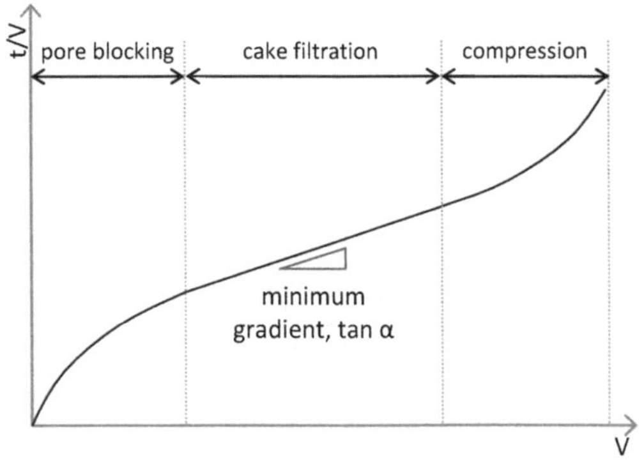

Figure 3-3 Filtration curve (t/V vs. V) and filtration mechanisms for constant pressure filtration

MFI can also be determined from the relation dt/dV versus V. The advantage of this approach is that the slope does not carry the effect of pore blocking as soon as cake/gel filtration occurs. Moreover pore blocking can be more accurately observed. A drawback is the need for very accurate monitoring of filtered volume as a function of time. Derivation and mathematical formulation of dt/dV versus V is presented by Boerlage (2001). The correction factor to be applied for MFI obtained from such curves is given by,

$$\mathrm{MFI} = \frac{1}{2}.\frac{\eta_{20°C}}{\eta_T}.\frac{\Delta P}{\Delta P_0}.\left(\frac{A}{A_0}\right)^2.\tan\alpha \qquad \text{(Eq. 3-6)}$$

3.2.2 Specific cake/gel resistance as a function of particle and cake properties

MFI depends on specific cake/gel resistance α and particle concentration, as it is directly proportional to the fouling index I,

$$I = \alpha.C_b \qquad \text{(Eq. 3-7)}$$

Where,
α is specific resistance of the deposited cake/gel layer (m/kg)
C_b is bulk particle concentration in the feed water (kg/m³).

Specific cake/gel resistance is related to particle properties, e.g., size, shape (sphericity), density, and cake/gel porosity and is given by the Carman-Kozeny relationship (Eq. 3-8). This relation shows that specific cake/gel resistance strongly depends on particle diameter and the porosity of the cake/gel layer,

$$\alpha = \frac{36\,K''}{\rho {d_p}^2 \Phi^3} \cdot \frac{(1-\varepsilon)}{\varepsilon^3} \qquad \text{(Eq. 3-8)}$$

Where,
K" is the Kozeny constant
ρ is particle density (kg/m^3)
d_p is particle diameter (m)
Φ is sphericity ($\Phi = 1$ for spherical particles)
ε is cake porosity.

The Kozeny constant K" is normally taken as 5 for porosity of packed impermeable spheres, i.e., 0.4 . However, it is very unlikely that this value is accurate where filtration is dependent on cake porosity, such as the case of flocculated aggregates. Nonetheless, the Carman-Kozeny equation can be used to provide insight into the effect of particle size on the specific cake resistance (Lee et al., 2003; Foley, 2013).

The porosity of a cake/gel layer formed from flocculated aggregates is a function of the density of the cake/gel layer. For cake/gel layers formed from porous aggregates, density per unit volume can be given by,

$$\rho_c = \varepsilon . \rho_w + (1-\varepsilon) . \rho_s \qquad \text{(Eq. 3-9)}$$

Where,
ρ_c is the density of cake/gel layer (kg/m^3)
ε is porosity of the cake/gel layer
ρ_w is the density of water (kg/m^3)
ρ_s is the density of primary particles of aluminium hydroxide (i.e., 2,420 kg/m^3).

If the density of aluminium hydroxide aggregates is known, the porosity of the cake/gel layer can be determined by Eq. 3-10. Density values for aluminium hydroxide aggregates are reported in the range of 1,010-1,750 kg/m^3 (Tambo and Watanabe, 1979; Hossain and Bache, 1991). In order to reflect a large range of porosities, i.e., 0.4-0.99, a range of densities from approximately 1,000 to 1,860 kg/m^3 were considered in this study.

$$\varepsilon = \frac{\rho_s - \rho_c}{\rho_s - \rho_w} \qquad \text{(Eq. 3-10)}$$

Specific cake resistance may be obtained from experimentally measured MFI values (Eq. 3-5). When particle concentration (C_b) is not accurately known, the fouling index (I) can be used to represent fouling potential of flocculated solutions.

In most cases it is not possible to determine C_b and α accurately. In this study however, C_b could be approximated as aluminium concentration retained on the membrane, by subtracting the residual aluminium measured in permeate samples from aluminium concentration dosed in the feed water. Particle concentration of tap water was assumed to be negligible.

3.2.3 Prediction of pressure development in MF/UF systems

In constant flux filtration, change in filtered volume over time is constant,

$$V = A.J.t \qquad \text{(Eq. 3-11)}$$

Assuming that cake/gel filtration is the dominant fouling mechanism in dead-end MF/UF systems, resistance of the cake/gel layer (Eq. 3-2) can be rewritten as,

$$R_c = I.J.t \qquad \text{(Eq. 3-12)}$$

Substituting Eq. 3-12 in Eq. 3-1 and rearranging to get pressure increase in constant flux filtration,

$$\Delta P = \eta.R_m.J + \eta.I.J^2.t \qquad \text{(Eq. 3-13)}$$

Pressure increase in constant flux filtration is proportional to the fouling index I and the square of filtration flux. At time $t = 0$, pressure difference across a membrane is a function of membrane properties (R_m), temperature and filtration flux. If particle concentration C_b is zero, no increase in pressure should be observed with time. Eq. 3-13 was used in this study to predict pressure increase in MF/UF membranes using theoretically and experimentally obtained I values.

Note: Particulate/colloidal fouling potential has been shown to depend - to a certain extent - on filtration flux and therefore MFI-UF constant flux has been developed (Boerlage et al., 2004; Salinas Rodriguez, 2011) and is now preferably applied, as more and more MF/UF systems are operated in constant flux in practice.

In this study predictions of pressure development in MF/UF operating at constant flux were made with I values obtained from measurements at constant pressure. This does not affect the main conclusion of this study since the flux in the experiments performed at constant pressure was much higher than the flux applied in MF/UF systems in practice.

3.3 Materials and methods

3.3.1 Feed water

To investigate the contribution of aluminium itself to fouling potential, feed water should be free of particles that may affect the structure and characteristics of aggregates e.g., suspended and colloidal particles and NOM. To this end, coagulation should ideally be carried out with ultra pure water to which inorganic compounds are added. However, from an experimental point of view (i.e., avoiding the introduction of foulants from salts) tap water, having a very low $MFI_{0.1}$ was preferred. Delft tap water was used as feed water for all experiments conducted in this study. The average feed water characteristics of the tap water were: pH 8.1; temperature 21 °C; alkalinity 2 mmol HCO_3/L; dissolved organic carbon (DOC) 1.9 mg C/L; ultraviolet absorbance (UV_{254}) 2.7 m^{-1}; turbidity < 1 NTU; and $MFI_{0.1}$ < 100 s/L^2.

3.3.2 Jar tests

Jar tests were conducted with a calibrated jar test unit. PACl (AquaRhône® 18D) with a basicity of 38 ± 5% was used as coagulant at doses of 1, 2 and 5 mg Al(III)/L, corresponding to 11.1, 22.2 and 55.6 mg/l as PACl respectively. PACl was chosen because it demonstrates a higher rate of floc formation than iron and aluminium salts (Gorczyca and Zhang, 2006), and is frequently used in inline coagulation applications for fresh surface water treatment. Several types of PACl are commercially available, differing in terms of basicity (ratio OH/Al) and Al content. Basicity ranges from 15 - 85%. In general, the higher the basicity, the larger the fraction of polymeric Al (Bottero and Cases, 1980). Temperature was recorded for each jar test and values ranged between 20-22 °C. pH was adjusted to values between 5.5 to 7.4-7.7 (depending on dose) by the addition of 1 M HCl and 1 M NaOH.

Flocculation was carried out in two different modes to study the effect of flocculation time on formation and characteristics of aluminium aggregates,

- Only rapid mixing for a very short time (10 seconds) at a G-value of 470 s^{-1}, to simulate inline coagulation (absence of flocculation).
- Rapid mixing for a period of 30 seconds at the same intensity as the first mode, followed by flocculation at 40 s^{-1} for different flocculation times (up to an hour), to investigate the effect of prolonged flocculation time on fouling potential of flocculated solutions.

3.3.3 Filtration experiments

Filtration tests were conducted using an unstirred cell device schematically presented in Figure 3-4. Batch experiments were performed using Amicon cells (8200 series) with a maximum process volume of 200 ml. The stirring assembly was removed from the cell to avoid the introduction of velocity gradient which might break up the aggregates or enhance their growth. Filtration was achieved under dead-end, constant pressure mode.

Flocculated suspensions prepared under different process conditions were transferred to the cell unit at the end of each flocculation time. The solution was filtered through a hydrophilic PVDF filter with a pore size of 0.1 μm (Millipore, USA) under constant pressure of 1 bar. Clean water flux was approximately 1,450 L/m^{2}h. Permeate from the cell was collected in a beaker set on an electronic balance (Mettler Toledo, Model PB 602-S). Data sets of collected filtrate weight and filtration time were recorded and processed to determine $MFI_{0.1}$.

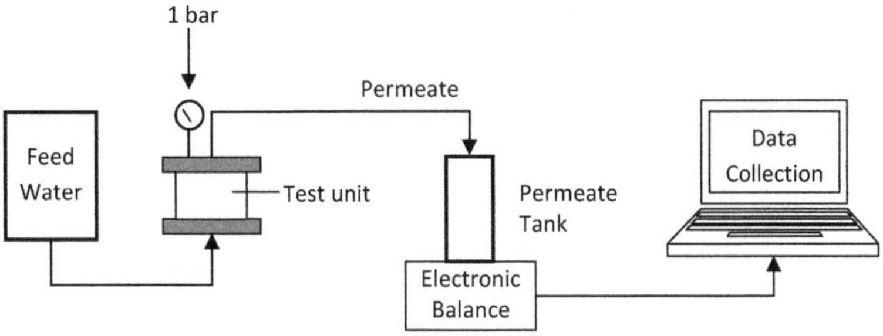

Figure 3-4 Schematic representation of the filtration setup

3.3.4 Analytical methods

For aluminium measurements, permeate samples were collected, preserved (by acidification with nitric acid to pH 0.3-0.5), stored and tested per Standard Methods for the Examination of Water and Wastewater (19th Ed.). All permeate samples were analyzed with Eriochrome Cyanine-R dye in accordance with APHA 3500 – Al D, using a spectrophotometer (Perkin Elmer, Lambda 20) for absorbance at 535 nm. Minimum aluminium concentration detectable by this method in the absence of fluorides and complex phosphates is approximately 6 μg/l.

3.4 Results and discussion

3.4.1 Pressure increase as a function of particle and cake properties

Assuming cake filtration is the predominant mechanism, pressure increase in one filtration cycle can be calculated based on MFI values using Eq. 3-13. Figure 3-5 shows pressure increase in MF/UF membranes for a range of MFI values at different filtration flux rates. Calculations were based on a filtration cycle of 1 hour at 20 °C. Particles were assumed to be spheres (Φ = 1) and sphericity was assumed to remain constant throughout filtration. MF/UF membranes are typically backwashed when pressure increase in one cycle is in the range of 0.4-1 bar, depending on manufacturer recommendations. Inside-out hollow fibre membranes are commonly backwashed when pressure increases in the range of 0.2-0.4 bar, while the threshold for outside-in hollow fibre membranes can reach 1 bar.

Calculations indicated that increase in pressure per filtration cycle at different MFI values is strongly dependent on filtration flux. Threshold pressure range (0.4-1 bar) for backwash corresponds with MFI values approximately in the range of 20,000 - 50,000 s/L^2 at 100 L/m^{2}h, while at 15 L/m^{2}h, MFI values much higher than 100,000 s/L^2 are required to cause the same pressure increase. In practice, filtration flux in the range of 15 - 100 L/m^{2}h is frequently applied.

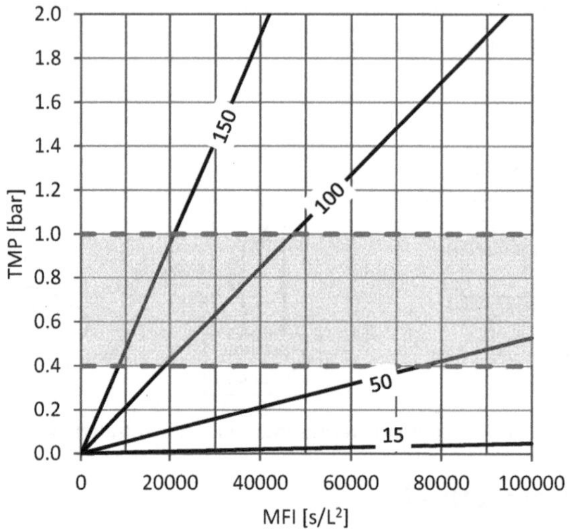

Figure 3-5 Theoretical pressure increase in MF/UF membranes as a function of MFI, in one filtration cycle of 1 hour at 20 °C for flux values of 15 - 150 L/m^{2}h. Dashed lines represent threshold pressure range for backwashing

Based on Eq. 3-5, Eq. 3-7 and Eq. 3-8, MFI values were calculated for a particle concentration of 1 mg/L and a range of particle size and cake porosity, for spherical particles ($\Phi = 1$). Cake porosity was calculated based on Eq. 3-10, assuming cake/gel density of up to 1,860 kg/m^3 (Figure 3-6). Resulting porosities range from 0.4-0.99; the lower limit reflects the porosity of sand, while porosity of cake/gel layers formed in membrane filtration falls at the upper limit of the calculated range.

Cake/gel layers of lower density, have higher porosity and lower MFI values. For the range of densities and porosities considered, particle size greater than 5 nm does not contribute significantly to fouling potential; e.g., for 5 nm particles forming a cake with a porosity of 0.4, the MFI value is approximately 50,000 s/L^2, which is at the defined threshold pressure for backwashing in MF/UF systems (Figure 3-6). Particles larger than 1 μm give MFI < 1 s/L^2 (data not shown in Figure 3-6).

Porosity has a large impact on MFI values. When cake porosity increases from 0.4 to 0.99, MFI is lower by more than two orders of magnitude for particle diameter of 5 nm. It is important to note that this is a simplified approach intended for illustration of the effect of particle size and cake porosity on specific cake resistance, and the assumptions in these calculations may not necessary reflect actual conditions during filtration of colloids and particles. Surface characteristics of the particles, change in sphericity due to compression, etc. can give different results than calculated. Aggregates formed from the flocculation of particles and colloids very often have a fractal, self-similar structure. An important consequence of this is that the effective density of aggregates decreases with increase in size (Gregory, 1997; Li and Logan, 2001) with large aggregates having much lower density than the primary particles from which they are made. The density-size dependency of aggregates was not taken into account in these calculations.

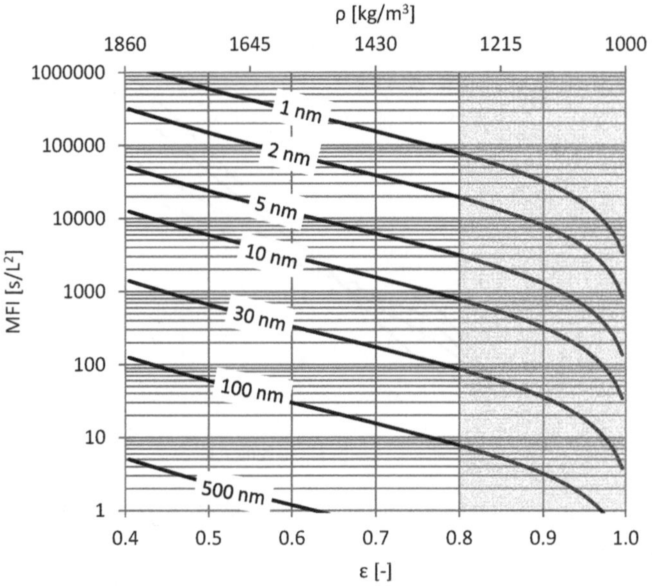

Figure 3-6 MFI values as a function of cake porosity for particle concentration of 1 mg/L, isolines represent particle size of primary particles. Shaded area represents porosity range for cake/gel layers in membrane filtration

These results indicate that in MF/UF systems, coagulation conditions can be chosen for the creation of relatively small flocs, providing the opportunity of minimizing or completely eliminating conventional flocculation.

3.4.2 Fouling potential of PACl in MF

3.4.2.1 Effect of dose and pH

Fouling potential of feed water coagulated at pH values of 5.5 to 7.4-7.7 and coagulant dose ranging from 1 to 5 mg Al(III)/L was evaluated. Rapid mixing was performed at 320 s^{-1} for 30 s followed by flocculation at 50 s^{-1} for 65 seconds (Gt 9600/3250). Figure 3-7 depicts the effect of pH on fouling potential as measured by MFI at constant flocculation Gt for different doses of aluminium.

At higher coagulant dose, higher MFI values were observed. This outcome was expected, as C_b increases with coagulant concentration. MFI values of up to 3,500 s/L² were obtained from the filtration of aluminium hydroxide flocs at varying coagulant dose and pH. These values are far below the lower threshold range of 20,000 s/L² (for filtration at 100 L/m²h) defined in Figure 3-5. Therefore the contribution of these aggregates to pressure increase in MF/UF systems, applying membranes with pore size larger than 0.1 μm, is expected to be marginal. At 1 mg Al(III)/L the effect of pH on MFI was the opposite of that at 5 mg Al(III)/L.

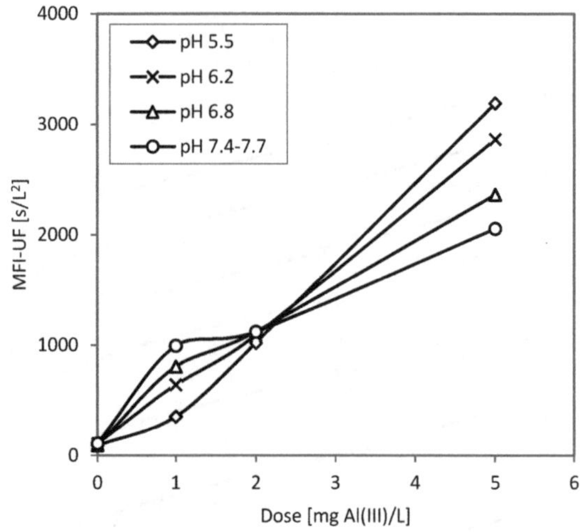

Figure 3-7 MFI as a function of dose and pH at Gt 9600/3250

In general the rate of flocculation is higher at higher pH. The remarkable difference in α values between 1 and 5 mg Al(III)/L might be attributed to flocculation kinetics as a function of pH at different coagulant dose. At 1 mg Al(III)/L and low pH, flocculation kinetics are slow, resulting in the formation of small particles that are not retained by 0.1 μm membrane filters. At higher pH, flocculation is enhanced, allowing for the formation of larger aggregates that are retained by the membrane filters, thereby imparting a higher resistance.

At 5 mg Al(III)/L, the effect of pH is reversed. For this coagulant dose, faster flocculation kinetics at high pH may result in enlargement of aggregates that form cake/gel layers with low specific resistance. The effect of pH on MFI and α is substantial in relative terms. However in absolute terms the level of MFI remains marginal.

3.4.2.2 Effect of Gt

Experiments were performed to compare the effect of Gt on fouling potential at different pH values for coagulation at 1 and 5 mg Al(III)/L. Details of the mixing conditions are presented in Table 3-1.

Table 3-1 Mixing time and intensity for different Gt values

	Rapid Mixing			Flocculation			
Exp. Set	G [s^{-1}]	t [s]	Gt	G [s^{-1}]	t [s]	Gt	Gt
I	320	30	9600	50	65	3250	9600/3250
II	470	30	14100	40	2100	84000	14100/84000

Figure 3-8 shows that MFI is lower at higher Gt values, independent from pH. This effect is attributed to the formation of larger flocs, having a lower specific resistance. The graphs are in concert with the observation presented in Figure 3-7, namely, at a dose of 1 mg Al(III)/L the effect of pH on α is the opposite of the effect at 5 mg Al(III)/L.

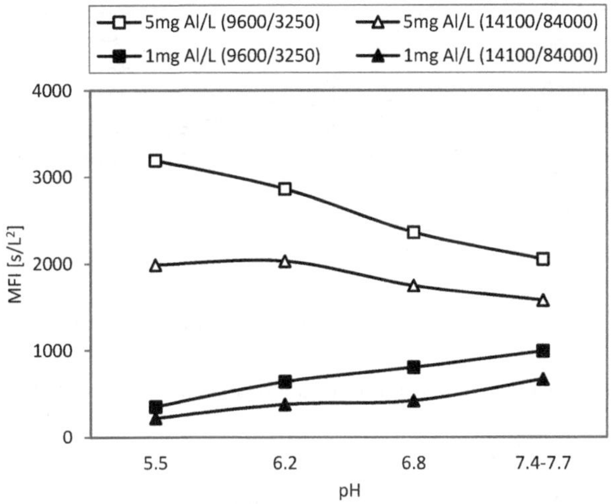

Figure 3-8 Effect of Gt and pH on fouling potential at 1 and 5 mg Al(III)/L

3.4.2.3 Effect of pH and Gt

The effect of Gt on formation and growth of aluminium aggregates was investigated at a constant dose of 5 mg Al(III)/L. The range of mixing time and intensities (rapid mixing and flocculation), used for this set of experiments are given in Table 3-2.

Table 3-2 Mixing time and intensities for assessing the effect of flocculation on fouling potential

	Rapid Mixing			Flocculation			
Exp. Set	G [s^{-1}]	t [s]	Gt RM	G [s^{-1}]	t [s]	Gt SM	Gt
I	470	10	4700	0	0	0	4700/0
	470	30	14100	40	578	23100	14100/23100
II	470	30	14100	40	2100	84000	14100/84000
	470	30	14100	40	3150	126000	14100/126000

pH values were stable after coagulant addition and mixing for the range of 5.5 to 7.4-7.7. For coagulation at 5 mg Al(III)/L, aggregates exposed to longer flocculation time, had lower fouling potential as measured by MFI (Figure 3-9). This effect was particularly noticeable for coagulation at pH 5.5, where flocculation kinetics is slower (Zhang et al., 2004). Under such conditions, prolonging flocculation time can result in further agglomeration and aggregate growth. As a consequence, lower MFI values are obtained. On the contrary, at pH 7.3, MFI did not change in function of flocculation time as precipitate formation occurs faster at this pH value. For flocculation times (4700/0 and 14100/23100), pH played a dominant role, whereby MFI was lower at high pH. This further confirms that pH affects aggregation kinetics.

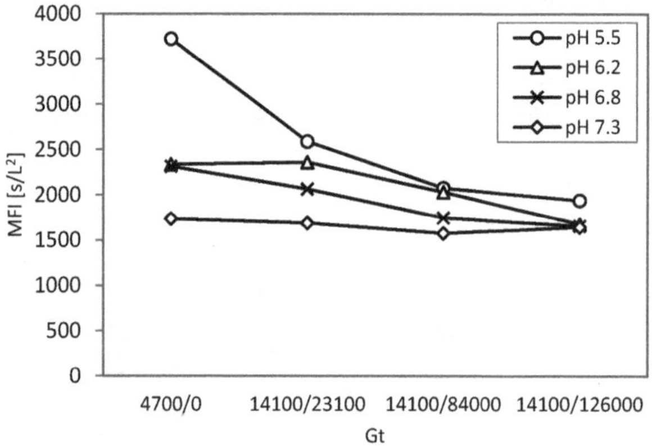

Figure 3-9 Effect of pH and flocculation Gt on MFI at 5 mg Al(III)/L

3.4.2.4 Aluminium residual

Figure 3-10 illustrates Al residual in permeate samples for feed water coagulated at 5 mg Al(III)/L at different Gt and pH values. At pH 5.5 and 7.4, higher passage was observed. Aluminium residual in the permeate samples at pH 6.2 and 6.8, were consistently lower than the two pH extremes. This is in good accordance with PACl solubility (Figure 3-11).

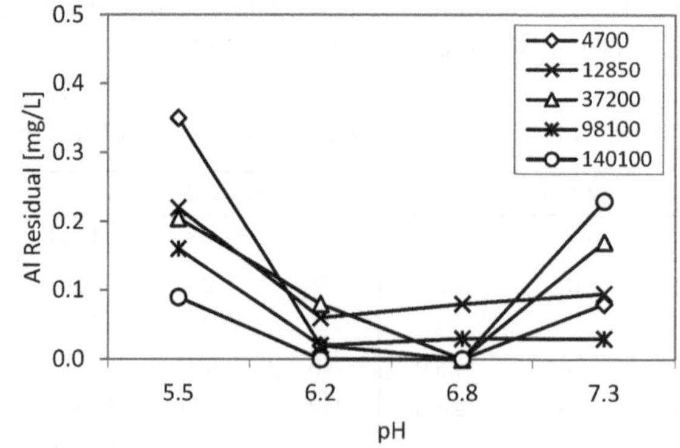

Figure 3-10 Residual aluminium as a function of pH and total Gt at coagulant dose of 5 mg Al(III)/L

At pH 5.5, increase in flocculation time, resulted in lower aluminium passage. Aluminium residual decreased from 0.35 mg/l at no flocculation to approximately 0.1 mg/l at 60 minutes flocculation. At pH 7.4, an opposite trend was observed; longer flocculation times resulted in a higher passage of aluminium. It may be that aluminium flocs formed at higher pH are weak and can rupture when exposed to high Gt values and pass through 0.1 μm pores. At lower pH values particles are dense, less porous and consequently less apt to breakage. Maximum aluminium passage is 0.35 mg/l and occurs at pH 5.5 when no flocculation is applied. This amounts to about 7% of the coagulant dose.

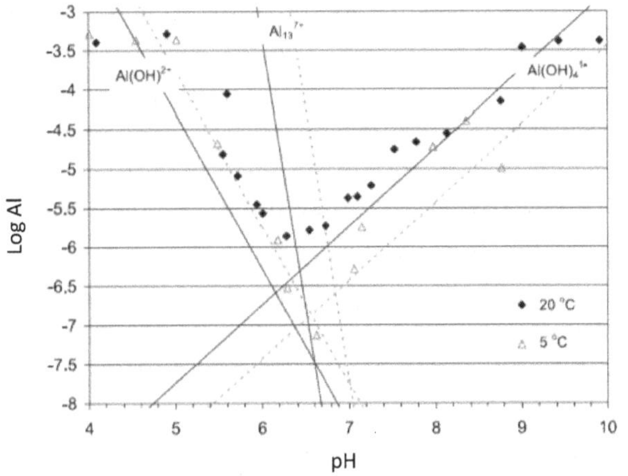

Figure 3-11 Experimental solubility data at 20 °C and 5 °C for PACl and theoretical solubility for Al species in equilibriums with $Al(OH)_{3(am)}$ indicated by solid (20 °C) and dashed (5 °C) lines (Pernitsky and Edzwald, 2003)

3.4.2.5 Filtration mechanism

Figure 3-12 shows results obtained from filtration of (a) tap water without coagulation at pH 6.8 and (b) coagulated tap water with 5 mg Al(III)/L, pH 6.8 and Gt 9600/3250. Pore blocking was initially observed for filtration with tap water without coagulant (Figure 3-11a). This trend was observed at all pH values.

When coagulant was added, pore blocking was no longer observed. MFI values obtained from filtration with tap water were constant at approximately 90 s/L^2 for the tested pH range. Results obtained from the filtration of coagulated solutions were assessed for pore blocking by (a) visual inspection of the filtration curves (dt/dV versus V curves); (b) re-calculation of initial membrane resistance from filtration results; and (c) evaluation of the slope of dt/dV versus V graphs. Pore blocking was not detectable by any of the mentioned approaches. It cannot be excluded that due to the high flux, the time span of the phenomenon was too short to be recorded at the very start of the filtration run.

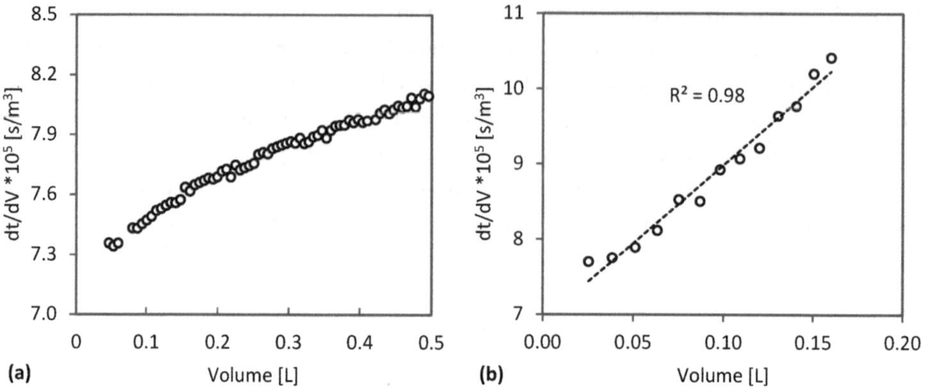

Figure 3-12 Filtration curves for (a) tap water without coagulation at pH 6.8 and (b) coagulated tap water at 5 mg Al(III)/L, pH 6.8 and Gt 9600/3250

3.4.3 Predicting pressure increase in MF/UF systems

Experimentally obtained MFI values were used to calculate pressure build-up in an MF/UF for a range of filtration flux rates according to Eq. 3-13. Results are illustrated in Figure 3-13. Pressure increase due to resistance imparted by aluminium flocs was less than 0.2 bar for one filtration cycle of one hour for almost all conditions tested. The effect of pH on specific cake resistance and pressure development was slightly more pronounced.

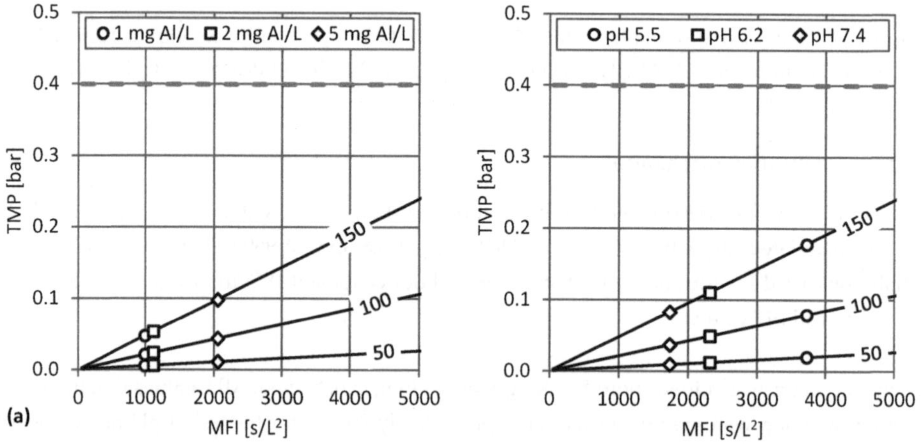

Figure 3-13 Calculated pressure development in MF/UF systems for 1 hour filtration time at different flux rates based on experimentally obtained MFI values from coagulation at (a) 1, 2 and 5 mg Al(III)/L, Gt 9600/3250 and 7.4-7.7, and (b) 5 mg Al(III)/L, Gt 9600/3250 as a function of pH; isolines represent theoretical pressure increase. Red dashed line represents lower pressure threshold for backwashing

However, the development of pressure remained marginal and well below the lower threshold level for backwashing in MF/UF systems. It may therefore be concluded that the contribution of aluminium flocs to pressure build-up in MF systems, applying membranes with pore size ≥ 0.1 μm, is marginal.

3.5 Conclusions

- Theoretical calculations indicate that
  - MFI levels below 20,000 s/L² result in a pressure increase of about 0.4 bar for one filtration cycle of 1 hour in MF/UF systems operating at 100 L/m²h. At a ten times lower flux of 10 L/m²h, MFI of 2,000,000 s/L² will give the same result.
  - Spherical primary particles of 5 nm diameter and larger - forming cake/gel layers with porosity ranging between 0.4 and 0.99 - do not contribute significantly to pressure increase in MF/UF systems operated in constant flux.
- Experimental results showed that the fouling potential of aluminium hydroxide aggregates - created under varying coagulant dose, pH, and flocculation time - in 0.1 μm membranes is low. A dose of 5 mg Al(III)/L for pH values ranging from 5.5 to 7.4-7.7 created a maximum MFI value of 3,500 s/L², resulting in a predicted pressure increase in the first filtration cycle of less than 0.08 bar at a flux of 100 L/m²h. From predictions of pressure increase in constant flux MF systems, it may be concluded that aluminium aggregates contribute a very small fraction to increase in pressure.
- The effect of pH on MFI of aluminium hydroxide aggregates at a dose of 5 mg Al(III)/L is in concert with expectations, namely, the higher the pH the lower the MFI. At a low dose of 1 mg Al(III)/L the opposite effect was observed, which is attributed to flocculation kinetics at low coagulant dose and pH.
- For filtration of tap water through 0.1 μm membranes, initial pore blocking was observed. When coagulant was added to the tap water pore blocking was no longer observed. The predominant filtration mechanism for solutions prepared under different flocculation conditions was cake/gel filtration from the very beginning of filtration.

3.6 Acknowledgements

Thanks are due to Aleid Diepeveen and Frederik Spenkelink for their technical support and contribution to this chapter.

3.7 References

Amirtharajah, A., Mills, K.M., 1982. Rapid-mix design for mechanisms of alum coagulation. Journal of the American Water Works Association 74 (4), 210-216.

Binnie, C., Kimber, M., Smethurst, G., 2002. Basic Water Treatment. 3rd ed. Cambridge: Royal Society of Chemistry. 59-71.

Boerlage, S.F.E., 2001. Scaling and particulate fouling in membrane filtration systems. Doctoral dissertation, IHE Delft/Wageningen University, The Netherlands.

Boerlage, S.F.E., Kennedy, M., Tarawneh, Z. Faber, R.D., Schippers, J.C., 2004. Development of the MFI-UF in constant flux filtration. Desalination 161, 103-113.

Bottero, J.Y., Cases, J.M., 1980. Studies of hydrolyzed aluminium chloride solutions. 1. Nature of aluminium species and composition of aqueous solutions. Physical Chemistry 84, 2933-2939.

Bratby, J., 2006. Coagulation and flocculation in water and wastewater treatment. Second Edition. London: IWA Publishing.

Brinkman, H.C., 1948. On the permeability of media consisting of closely packed porous particles. Journal of Applied Sciences Research 1, 81-86.

Camp, T.R, Stein, P.C., 1943. Velocity gradients and internal work in fluid motion. Journal of the Boston Society of Civil Engineers, October.

Camp, T.R., 1955. Flocculation and flocculation basins. Transactions, American Society of Civil Engineers 120, 1-16.

EPA, USEPA, 1999. Enhanced coagulation and enhanced precipitative softening guidance manual. Disinfectants and disinfection byproducts rule (DBPR), 237.

Foley,G., 2013. Membrane filtration: A problem solving approach with MATLAB. Edinburgh: Cambridge University Press.

Gabelich, C., Ishida, K.P., Gerringer, F.W., Evangelista, R., Kalyan, M., Suffet, I.H., 2006. Control of residual aluminium from conventional pretreatment to improve reverse osmosis. Desalination 190, 147-160.

Gorczyca B., Zhang G., 2006. Solid/liquid separation behaviour of alum and polyaluminium chloride coagulations flocs. Water Science Technology: Water Supply 6, 79-86.

Gregory, J., 1997. The density of particle aggregates. Water Science and Technology 36 (4), 1-13.

Guigui, C., Rouch, J.C., Durand-Bourlier, L., Bonnelye, V., Aptel, P., 2002. Impact of coagulation conditions on the inline coagulation/UF process for drinking water production. Desalination 147, 95-100.

Hahn, H.H., Stumm, W., 1968. Kinetics of coagulation with hydrolysed Al (III). Journal of Colloid Interface Science 28, 134-144.

Hossain, M.D., Bache, D.H., 1991. Composition of alum flocs derived from a coloured low-turbidity water. Journal of Water Supply: Research and Technology-AQUA 40, 298-303.

Hwang, K.-J., Liu, H.-C., 2002. Cross-flow microfiltration of aggregated submicron particles. Journal of Membrane Science 201, 137–148.

Lee, S.A., Fane, A.G., Amal, R., Waite, T.D., 2003. The effect of floc size and structure on specific cake resistance and compressibility in dead-end microfiltration. Separation Science and Technology 38 (4), 869-887.

Letterman, R.D., Quon, J.E., Gemmell, R.S., 1973. Influence of rapid-mix parameters on flocculation. Journal AWWA 65, 716-722.

Li, X, Logan, B.E., 2001. Permeability of fractal aggregates. Water Research 35 (14), 3373-3380.

Pernitsky, D.J., Edzwald, J.K., 2003. Solubility of polyaluminium chloride. Journal of Water Supply: Research and Technology-AQUA 52 (6), 395-406.

Ruth, B.F., Montillon, G.H., Montana, R.E., 1933. Studies in filtration. Industrial and Engineering Chemistry 25 (1), 76-82.

Salinas Rodriguez, S.G., 2011. Particulate and organic matter fouling of SWRO systems: characterization, modelling and applications. Doctoral dissertation, UNESCO-IHE/TU Delft, Delft.

Schippers, J.C., Verdouw, J., 1980. The modified fouling index, a method of determining the fouling characteristics of water. Desalination 32, 137-148.

Schurer R., Tabatabai A., Villacorte L., Schippers J.C., Kennedy M.D., 2013. Three years operational experience with ultrafiltration as SWRO pretreatment during algal bloom. Desalination and Water Treatment 51(4-6), 1034-1042.

Schurer, R., Janssen, A., Villacorte, L., Kennedy, M.D., 2012. Performance of ultrafiltration & coagulation in an UF-RO seawater desalination demonstration plant. Desalination and Water Treatment 42(1-3), 57-64.

Tambo N., Watanabe Y., 1979. Physical characteristics of flocs - I. The floc density function and aluminum floc. Water Research 13, 409-419.

van Benschoten, J.E., Edzwald, J.K., 1990. Chemical aspects of coagulation using aluminium salts - I. Hydrolytic reactions of alum and polyaluminium chloride. Water Research 24 (12), 1519-1526.

Ventresque, C., Gisclon, V., Bablon, G., Chagneau, G., 2000. An outstanding feat of modern technology: the Mery-sur-Oise nanofiltration treatment plant (340,000 m^3/d). Desalination 131, 1-16.

Zhang, P., Hahn, H.H., Hoffmann, E., 2004. Study on flocculation kinetics of silica particle suspensions. In: Hahn, H.H., Hoffmann, E., Odegaard, H. (Eds.) Chemical water and wastewater treatment. First Edition. Florida: IWA Publishing.

4

OPTIMIZING INLINE COAGULATION IN MICRO- AND ULTRAFILTRATION OF SURFACE WATER

This chapter reports on explorative investigations on the application of ferric chloride as coagulant in micro- and ultrafiltration systems (MF/UF). Investigations were conducted on the effect of process conditions during coagulation, e.g., dose, pH, mixing (G) and flocculation (Gt) conditions. Fouling potential was measured as MFI, using membrane filters of 0.1 μm, 300 kDa and 100 kDa. Three different water sources were investigated, Delft Canal water (fresh water); North Sea water (real seawater); synthetic seawater spiked with – laboratory prepared – algal organic matter. Due to the low fouling potential of fresh water and real seawater, coagulation resulted in higher MFI under almost all tested conditions. In synthetic seawater spiked with algal organic matter, coagulation demonstrated a very pronounced improvement of the fouling potential. Flocculation after rapid mixing resulted in some to marginal improvement of the fouling potential. Calculations showed very high G-values and short residence time prior to and within MF/UF elements, which are likely sufficient in achieving and/or maintaining relatively low fouling potential of feed water reaching the membrane surface. Observations in this chapter suggest that the main mechanism involved in fouling control in inline coagulation is most likely the improvement of backwashability of the fouling layer.

This chapter is based on parts of:

Tabatabai S.A.A., Gaulinger S.I., Kennedy M.D., Amy G.L., Schippers J.C., 2009. Optimization of inline coagulation in integrated membrane systems: a study of $FeCl_3$. Desalination and Water Treatment 10, 121-127.

Tabatabai S.A.A., Hernandez Caballero M., Hassan A., Ghebremichael K., Kennedy M.D., Schippers J.C., 2011. Optimizing coagulation in seawater UF/RO to reduce fouling by transparent exopolymer particles (TEPs). Proceedings of IDA World Congress, Perth.

4.1 Introduction

Since the mid 1980s, large scale application of micro- and ultrafiltration (MF/UF) in drinking water production has increased considerably. One of the main drivers behind this growth has been the implementation of stringent legislation for disinfection of drinking water in Europe and USA, requiring the use of advanced technology including MF/UF. Other important triggers behind the recent surge in MF/UF growth have been the development of capillary MF/UF membranes and the change from cross-flow to dead-end filtration. Adopting dead-end filtration in large-scale MF/UF plants has reduced energy consumption considerably making this technology more competitive with conventional treatment processes.

MF/UF membranes are employed for pretreatment of surface water and secondary treated wastewater effluent prior to nanofiltration (NF) and/or reverse osmosis (RO) (Ventresque et al., 2000). More recently, growth in the application of seawater desalination for drinking and industrial water supply has triggered the use of MF/UF membranes as pretreatment for seawater reverse osmosis (SWRO) membranes (Wolf et al., 2005; Halpern et al., 2005; Busch et al., 2010).

MF/UF are attractive alternatives to conventional treatment schemes as variations in feed water quality do not affect product water quality for a given water matrix. In order to ensure the competitiveness of MF/UF applications in terms of productivity, high fluxes of up to 100 L/m^2h are a pre-condition. However, a rapid reduction in permeability at these high operational fluxes might occur under severe fouling conditions, resulting in the need to apply very frequent backwashing and chemically enhanced backwashing (CEB). In addition, cleaning in place (CIP) may be needed as well. These conditions result in more complicated operation, lower recovery, longer downtime, higher operational costs and insufficient production capacity.

To reduce the frequency of backwashing and CEB for the control of reversible and irreversible fouling in MF/UF, two options are commonly applied,

- operation at lower flux;
- coagulant application - mainly in inline mode.

Inline coagulation is a process whereby coagulant is added to feed water prior to membrane filtration, but unlike conventional coagulation, coagulated solids are not removed by sedimentation or flotation. Inline coagulation is attractive, since higher fluxes can be maintained with a fairly simple process scheme, i.e., coagulant dosing and mixing in front of the MF/UF system. Successful implementation of inline coagulation in MF/UF systems in practice indicates that large flocs - as required in conventional systems - are not needed to control membrane fouling. This eliminates the need for extensive flocculation. The absence of flocculation and clarification steps results in lower investment cost for inline coagulation compared to conventional coagulation (coagulation, flocculation followed by sedimentation/flotation) applied in e.g., drinking water production.

Both aluminium and ferric coagulants are commonly used. However, when MF/UF membranes are used as pretreatment to NF/RO, ferric is applied, as aluminium precipitation can cause severe fouling problems in NF/RO systems (Ventresque et al., 2000; Gabelich et al., 2006).

This chapter is based on inline coagulation with ferric chloride prior to capillary MF/UF membranes operated in inside-out mode. In practice, rather high coagulant doses of up to 10 mg Fe(III)/L may be encountered in these systems (Lattemann, 2010; Edzwald and Haarhoff, 2011). When membrane filtration is combined with inline coagulation, backwash water containing iron residual has to be treated in many countries. Coagulation sludge requires concentrating, dewatering and disposal to comply with legislation in many parts of the world. These components substantially add to the cost and complexity of the process (Pearce, 2011). The challenge is therefore to optimise and simplify inline coagulation by minimizing coagulant consumption. In optimizing inline coagulation, process conditions such as dose, pH, mixing (G) and flocculation (G and Gt) might play a dominant role.

The performance of MF/UF systems may be characterized by four main aspects,

- Pressure development in subsequent filtration cycles;
- Permeability restoration by backwashing;
- Permeability restoration by chemically enhanced backwashing;
- Permeability restoration by cleaning in place.

The main focus of this chapter was on the first aspect by quantifying fouling potential in one filtration cycle in terms of Modified Fouling Index (MFI) measured with membranes of different pore sizes. Three water types were studied, namely,

- Low salinity surface water;
- North Sea water;
- Synthetic seawater spiked with algal organic matter (AOM) produced by laboratory-cultivated algae.

In addition an effort was made to get an indication of backwashability and fouling mechanisms in coagulation followed by MF/UF.

4.2 Background

4.2.1 Flocculation

After coagulant dispersion and partial colloid destabilization is achieved by rapid mixing, flocculation is applied by gentle stirring of the liquid to accelerate the rate of particle collisions and promote agglomeration of destabilized particles. Particle agglomeration results in reduction of the number of primary particles and enlargement of particle/aggregate size. Two main mechanisms involved in flocculation are perikinetic and orthokinetic aggregation.

4.2.1.1 Perikinetic aggregation

For submicron particles, transport rate is driven by Brownian motion. Von Smoluchowski (1917) developed the framework for modelling perikinetic aggregation where the frequency of collisions was obtained from the diffusional flux of particles toward a single stationary particle.

$$-\frac{dn}{dt} = \frac{4.E.k_B.T.n^2}{3.\eta} \qquad \text{(Eq. 4-1)}$$

Where,
E is collision efficiency factor (number of collisions that result in aggregation)
k_B is the Boltzmann constant (1.38×10^{-23} J/K)
T is absolute temperature (K)
n is the number of particles per unit volume
η is absolute viscosity (Ns/m^2).

4.2.1.2 Orthokinetic aggregation

Orthokinetic aggregation is more effective for larger aggregates whose motion is not governed by diffusion. In this process the collision frequency of particles/aggregates is enhanced by shear-induced velocity gradients.

The Von Smoluchowski theory of orthokinetic flocculation is applicable to a laminar flow field that has a well defined velocity gradient. Camp and Stein (1943) extended the use of this theory for application to turbulent fields by defining a measurable average value termed the 'root mean square velocity gradient (G) which is defined as,

$$G = \sqrt{\frac{P}{\eta.V}} \qquad \text{(Eq. 4-2)}$$

Where,
P is total power dissipated (J/s)
V is volume of fluid (m^3)
η is absolute viscosity (Ns/m^2).

Based on the concept of average velocity gradient (G), rate of particle agglomeration for orthokinetic flocculation is given by,

$$-\frac{dn}{dt} = \frac{4.E.G.\varphi_i.n}{\pi} \qquad \text{(Eq. 4-3)}$$

Where,
G is average velocity gradient (s^{-1})
φ_i is volume fraction of particles.

4.2.1.3 Flocculation in practice

Conventional water treatment schemes (e.g., coagulation followed by sedimentation and media filtration) are oriented toward the formation of flocs that are effectively removed by settling. As such, flocculation is designed for the creation of large flocs, e.g., few hundred microns to few millimetres, with high settling velocities. For design and operation of these systems, the parameters G and Gt are commonly used. Flocculators are usually designed for G-values in the range of 10 to 60 s^{-1} and Gt values in the range of 10,000 to 100,000 for conventional water treatment schemes (Bratby, 2006). Camp (1955) suggested G-values of 20 to 74 s^{-1} and flocculation time between 20 and 45 minutes resulting in Gt values in the range of 20,000 - 200,000 for flocculation.

For treatment of fresh surface water having relatively low turbidity, algal cell count and natural colour, inline coagulation in combination with media filtration is commonly applied. These systems do not comprise flocculation units and flocculation is expected to occur in the filter bed - due to elevated G-values resulting from the head loss across the filter bed - enhancing deep bed filtration.

In MF/UF applications, coagulation prior to membrane filtration is applied for fouling mitigation (Leiknes, 2009). Inline coagulation in MF/UF systems is generally achieved by dosing coagulant upstream of an inline mixer or the feed pump of the MF/UF membranes, and is characterized by very short residence time for mixing and the absence of a flocculation chamber. In these systems, the main aim of coagulation is to create particles which form cake/gel layers with high permeability and to "neutralize" particles responsible for non-backwashable fouling of MF/UF membranes.

Based on the Carman-Kozeny relation, resistance of a cake/gel layer is related to aggregate size and cake/gel layer porosity. The larger the aggregates, the higher the porosity and consequently higher permeability, i.e., lower resistance to filtration (See Chapter 3, section 3.2.2). Therefore, coagulation should be designed for the formation of aggregates of a certain size so as to enhance cake/gel layer permeability and the complexation, adsorption and/or enmeshment of particles in such a way that backwashability is improved.

Calculations based on the Carman - Kozeny equation indicated that coagulated particles in the micrometre range and even smaller agglomerate to form aggregates with high porosity resulting in the creation of cake/gel layers with low specific cake resistance 'α', i.e., high permeability (See Chapter 3, Figure 3-6).

This finding suggests that guidelines for flocculation in conventional treatment schemes may not be applicable for inline coagulation in MF/UF systems and much higher G-values in combination with short flocculation times may be applied.

4.2.2 Coagulation of organic matter

Collision efficiency 'E' plays a dominant role in agglomeration of hydrophobic particles (O' Melia and Stumm, 1967). If conditions are very unfavourable, i.e., presence of strong repulsive forces and/or absence of attractive forces, E is close to 0, whereas if no energy barrier is present and attractive forces dominate E approaches unity. Coagulants improve collision efficiency by,

- Charge neutralization;
- Inter-particle bridging;
- Enmeshment (during sweep flocculation).

Destabilization of hydrophilic colloids does not occur by compression of the electric double layer (as is the case for inorganic particles, e.g., clay), but rather these colloids are able to form soluble and insoluble complexes with metal ions (Gregor et al., 1997). Inter-particle bridging might be a contributing mechanism (Bache et al., 1999).

The negative charge of hydrophilic colloids, e.g., humic and fulvic acids, in natural waters is primarily due to the ionization of ionic groups such as carboxylic, aliphatic and aromatic hydroxyl, sulfato, phosphato, and amino groups, that functionalize the surface of these particles. Other negatively charged colloids such as polysaccharide gums and mucoproteins contain ionized carboxylic groups (Stumm and Morgan, 1962). Complex formation with metal ions might be accompanied by a reduction of the mean charge of these particles and by an alteration of their solubility (Stumm and Morgan, 1962). The capacity of these compounds in forming complexes with metals varies remarkably for different functional groups. This is attributed to the large variation in the affinity of inorganic cations toward the different functional groups (Pivokonsky et al., 2006).

The mechanisms involved in coagulation of organic matter are assumed to be complexation by precipitation of positively charged coagulant species and negatively charged functional groups on hydrophilic particles forming multi-complexes (Stumm and Morgan, 1962; Bratby, 2006), and adsorption of organic matter on precipitated metal hydroxide during sweep floc (Shen and Dempsey, 1998; Bache et al., 1999). It has been shown that coagulation of hydrophilic colloids is strongly influenced by the pH of the medium and the composition of the ionized groups (Stumm and Morgan, 1962).

Algal organic matter (AOM) has been linked to operational problems in MF/UF and RO systems. A major component of AOM in fresh and seawater is composed of transparent exopolymer particles (TEP). TEP have been identified as a major culprit in fouling of MF/UF membranes and potentially RO membranes in seawater desalination (Villacorte et al., 2010; Berman et al., 2011). These sticky particles, that are abundant in fresh and saline surface waters, have long been overlooked as a foulant. TEP are negatively charged, hydrophilic acidic polysaccharides and glycoproteins that could be up to a few hundred micrometres in length (Passow and Alldredge, 1994). The stickiness of TEP has been attributed to the presence of sulfate half-ester groups (Zhou et al., 1998). These particles are transparent (invisible) but can be visualized by staining with Alcian blue (Alldredge et al., 1993; Passow and Alldredge, 1995). A semi-quantitative method for the determination of TEP concentrations based on staining with Alcian blue has been developed very recently by Villacorte (2014). The Liquid Chromatography - Organic Carbon Detection (LC-OCD) method developed by DOC-Labor provides information on the presence of biopolymers smaller than 2 μm (Huber and Frimmel, 1994; Huber et al., 2011).

The mechanism of interaction between coagulant and TEP - ranging in size from several hundred micrometres down to several hundred nanometres - is assumed to be similar to humic acids and fulvic acids. At the same time, inter-particle bridging might occur as well (Bernhardt et al., 1985).

4.2.3 Velocity gradients

G and Gt values introduced in inline coagulation combined with MF/UF (operated in inside-out mode) are generated due to,

- Rapid mixing of coagulant;
- Pumping stage;
- Head loss in headers and connecting pipes;
- Head loss inside capillary membranes.

Rapid mixing and transfer through a pump is not expected to contribute to flocculation due to very high G-values and very short residence time. In conventional systems, high energy inputs or G-values during flocculation may cause break-up or disaggregation of the formed aggregates. In inline coagulation applications, where a flocculation stage is not formally incorporated in the process scheme, flocculation might only occur during transport from the pump to MF/UF membranes and within the MF/UF elements.

G-values can be calculated based on head loss in a pipe/conduit or an MF/UF capillary for a given flow rate as,

$$G = \left(\frac{Q\Delta P}{\eta V}\right)^{1/2} \qquad \text{(Eq. 4-4)}$$

Where,
Q is volumetric flow (m^3/s)
ΔP is head loss between two ends of a pipe (Pa).

To determine head loss or pressure drop between the two ends of the pipe or capillary the Hagen-Poiseuille law can be applied,

$$\Delta P = \frac{128\eta LQ}{\pi D^4} \qquad \text{(Eq. 4-5)}$$

Where,
L is length of the pipe (m)
D is pipe diameter (m).

Substituting Eq. 4-5 in Eq. 4-4 and rearranging,

$$G = \left(\frac{128\eta LQ^2}{\eta\pi D^4 V}\right)^{1/2} = \left(\frac{128QL}{\pi D^4 t}\right)^{1/2} \qquad \text{(Eq. 4-6)}$$

4.3 Materials and methods

4.3.1 Feed water

Low salinity surface water: Delft Westvest canal water pre-filtered through a 122 μm sieve was used for the experiments. Average feed water characteristics of the canal water were pH 7.9; total organic carbon (TOC) 18.4 mg C/L; dissolved organic carbon (DOC) 16 mg C/L; TSS 12.6 mg/L; and Fe 0.3 mg/L. When preparing a test series, pre-filtered canal water was taken from the batch container stored at 4°C and allowed to adjust to room temperature (20 °C). Before the sample was taken, the container was thoroughly stirred in order to avoid heterogeneity in suspended matter content for different tests.

North Sea water: North Sea water - pre-strained through 50 μm - was collected from a seawater UF/RO pilot desalination plant in Zeeland province, the Netherlands (sampling date, May 2010). Samples were collected in clean, dark-glass bottles and stored at 4 °C. Experiments were performed in less than a week after sample collection. Before each experiment, bottles were shaken gently to ensure that any particulate matter was brought back to suspension. Feed water quality parameters were pH 8; turbidity 5.7 NTU; EC 4800 mS/m; and TOC 1.8 mg C/L. $TEP_{0.4}$ at the time of sampling was approximately 0.093 mg Xeq/L, which is low as compared to absorbance values during algal bloom conditions (Villacorte, 2014).

Synthetic seawater spiked with algal organic matter: The fouling potential of seawater taken from the intake of the pilot plant turned out to be very low, most likely due to the absence of algae at the moment of sampling. To overcome this disadvantage, solutions of algal organic matter (AOM) in synthetic seawater were used. Synthetic seawater (SSW) was prepared following the recipe of Lyman and Fleming (1948) with a TDS of 35 g/L and electric conductivity (EC) of approximately 48,000 μS/cm.

AOM was isolated from a marine diatom species, *Chaetoceros affinis*. The diatoms were introduced into a previously autoclaved system with synthetic seawater as medium and allowed to grow under constant light conditions at room temperature (20°C) for up to 14 days. Nutrient supply followed the recipe of f/2+Si medium for marine diatoms (CCAP, 2010). The cultures were set on a moderate shaking speed (~ 100 rpm) to avoid settling of the cells. Openings were sealed with cotton to allow for air circulation. After the defined period, cultures were let to settle for 24 hours.

To quantify AOM concentration in the solutions, samples were analyzed with liquid chromatography – organic carbon detection (LC-OCD) by DOC-Labor (Karlsruhe, Germany). The biopolymer concentration obtained from the LC-OCD analysis was considered as an indication of the concentration of AOM; expressed in mg C/L as biopolymers. Feed solutions for experiments were prepared by diluting AOM stock solutions to a concentration of 0.7 mg C/L as biopolymers, representing the upper range of measured biopolymer concentrations in North Sea water from the Oosterschelde during spring algal bloom.

4.3.2 Jar tests

Low salinity surface water: Coagulation was done in batch mode using a calibrated jar test unit equipped with a continuously adjustable speed regulator. Rotation speed of the flat-bladed impellers was measured with a digital contact tachometer (Model 461891, Extech Instruments). Velocity gradients were obtained from calibration graphs of G-values versus rotation speed (rpm). Feed water pH was adjusted prior to addition of coagulant ($FeCl_3$) with 1 M HCl and 1 M NaOH. Rapid mixing intensity was set to 470 s^{-1} for 20 to 30 seconds, followed by flocculation at an intensity of 35 s^{-1} for durations of 2 to 60 minutes.

High salinity surface water: For experiments with North Sea water and synthetic seawater spiked with AOM, coagulation was done in batch mode with a calibrated jar test unit (Model CLM4, EC Engineering) with $FeCl_3$. pH adjustment was done with 1 M HCl or 1 M NaOH as necessary. Two sets of coagulation tests were carried out. In the first set, only rapid mixing was applied for 20 seconds at an intensity of 1,100 s^{-1} to simulate inline coagulation. In the second set rapid mixing was followed by flocculation at different G and Gt values.

Note: Gt values imparted during transfer of coagulated samples from jar tester to filtration cell are not accounted for. However, care was taken to avoid introduction of G-values and time, by immediate and gentle transfer of samples to the filtration cell.

4.3.3 MFI measurements

Theoretical background on the development of MFI is discussed extensively in Chapter 3, section 3.2.1. After completion of the targeted flocculation time, coagulated suspensions were immediately transferred to the filtration setup with care, to avoid breakage of flocs. For all tests an unstirred cell (Amicon Series 8000, Millipore) was used. The magnetic stirring mechanism was removed from the cell to avoid floc growth or breakage. Filtration was achieved at constant pressure in dead-end mode. Applied pressure, filtrate volume and membrane specifications are given in Table 4-1.

Filtrate was collected in a beaker placed on a digital scale (Model PB 602-S, Mettler Toledo). The scale had an RS-232 interface with a computer. Data sets of filtrate weight collected over time were recorded and imported into an MS Excel spreadsheet by data acquisition software (WinWedge, TALtech) for further processing. For each test run the membrane was flushed with 1 litre of deionised water with a resistivity of 18.2 MΩ.cm to measure clean water permeability at the same pressure as that used for subsequent filtration of feed samples.

Table 4-1 Experimental conditions for MFI measurements for different feed waters

Feed water type	Applied pressure [bar]	Cell volume [mL]	MWCO	Membrane area [cm^2]	Membrane material	Clean water permeability [L/m^{2}hbar]
Low salinity surface water	1	200	0.1 μm	27.8	PVDF*	1,450
North sea water	1	400	300 kDa	41.8	PES+	950
SSW spiked with AOM	0.3	400	100 kDa	41.8	PES+	1,300

* Millipore, USA
+ Omega, Pall Corporation, USA

4.3.4 Pore blocking and permeability restoration

To get an indication of the contribution of pore blocking to membrane fouling, the cake/gel layer was removed from the membrane surface with a soft sponge followed by rinsing the membrane with Milli-Q water. Thereafter Milli-Q water was filtered through the cleaned membranes and membrane resistance was measured. The difference between initial and final membrane resistance was attributed to pore blocking.

4.3.5 Analytical methods

4.3.5.1 DOC, UV_{254}, Iron

Organic matter characterization of fresh water samples was done by measuring DOC concentration, UV_{254} absorbance and calculating specific ultraviolet absorbance (SUVA). DOC was measured with a Shimadzu TOC analyzer. UV_{254} was measured using a UV-2501 PC spectrophotometer (Shimadzu, USA). SUVA was calculated by normalizing UV_{254} absorbance values for DOC concentration and expressed in L/mg.m.

Iron residual in the permeate samples was measured with inductively coupled plasma (ICP).

4.3.5.2 LC-OCD

LC-OCD was used for quantification and fractionation of organic carbon in solutions of AOM (containing TEP) in SSW. LC-OCD is based on size-exclusion chromatography in combination with organic carbon and organic nitrogen detection (Huber et al., 2011). Prior to separation, samples are made particle-free by filtration through 0.45 μm filters. To determine AOM concentrations, the 0.45 μm filter was bypassed upon request, resulting in a chromatographable limit of 2 μm. This limit is dictated by the interstitial voids in the column, and the frits at the entrance and exit of the column (Huber, pers. comm.). It may not be excluded that part of the sticky AOM (< 2 μm) may remain in the column due to charge interactions or sticking.

4.4 Results and discussion

4.4.1 Low salinity surface water

4.4.1.1 Effect of dose and Gt

The influence of flocculation time (ranging from 0 to 60 minutes), and coagulant dose (ranging from 0 to 15 mg Fe(III)/L) on fouling potential - measured as $MFI_{0.1}$ - of pre-filtered canal water was assessed. Feed water pH was adjusted prior to coagulation to achieve a final value of 8. Filtration was performed at a pressure of 1 bar at 21°C. Details of mixing and flocculation conditions for the experiments are presented in Table 4-2.

Table 4-2 Mixing and flocculation conditions for coagulation with $FeCl_3$, at pH 8

Exp.	Rapid Mixing			Flocculation			Gt
	G [s^{-1}]	t [s]	Gt	G [s^{-1}]	t [min]	Gt	
a	470	20	9400	0	0	0	9400/0
b	470	30	14100	35	2	4200	14100/4200
c	470	30	14100	35	11	23100	14100/23100
d	470	30	14100	35	30	63000	14100/63000
e	470	30	14100	35	40	84000	14100/84000
f	470	30	14100	35	60	126000	14100/126000

Delft canal water had a fouling potential - as measured by $MFI_{0.1}$ - in the range of 700 s/L^2 (Figure 4-1). This is relatively low for surface water without pretreatment. Based on theoretical calculations, this MFI value corresponds to a pressure increase of approximately 0.01 bar in one filtration cycle of 1 hour in an MF/UF system operated at 100 L/m^2h (Refer Chapter 3, Figure 3-5).

Addition of coagulant resulted in an increase in $MFI_{0.1}$ of the feed solutions for almost all conditions. However, the $MFI_{0.1}$ values are still in the low range for MF/UF operation and theoretically, the coagulated solutions will not contribute significantly to pressure increase in one filtration cycle.

At 0.5 mg Fe(III)/L and pH 8, the concentration of positively charged iron species is relatively low. This might result in the formation of colloidal iron species and/or colloidal Fe-NOM compounds. By definition, small particle size and/or cake porosity result in high MFI values. Consequently, colloidal iron species, with or without NOM, may form cake/gel layers with low porosity and high $MFI_{0.1}$. At high coagulant dose, particle concentration (C_b) is increased, resulting in higher $MFI_{0.1}$ (for theoretical background and related equations on MFI, refer to Chapter 3, section 3.2.1 and 3.2.2).

For all coagulated samples, a minimum $MFI_{0.1}$ was observed at 1 mg Fe(III)/L, beyond which increase in coagulant dose resulted in higher $MFI_{0.1}$ values. The predominant coagulation mechanism is assumed to be sweep floc, whereby organic matter is adsorbed to and enmeshed in precipitated iron hydroxide. Complexation reactions may only play a limited role, as the proportion of positively charged iron species is relatively small at pH 8. At 1 mg Fe(III)/L, Fe-NOM aggregates form a cake/gel layer with high porosity and consequently high permeability (low $MFI_{0.1}$). The high permeability of the cake/gel layer compensates in part for the additional resistance imparted by the coagulated solids (Fe-NOM aggregates and iron hydroxide precipitates). As coagulant dose is increased, larger concentrations of coagulated solids are retained by the membrane and higher $MFI_{0.1}$ values are observed, since MFI is proportional to particle concentration C_b.

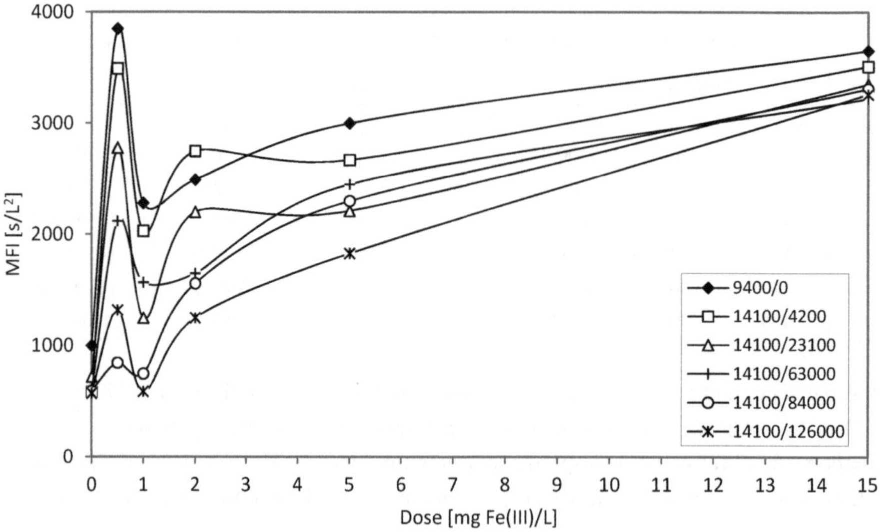

Figure 4-1 Fouling potential of Delft canal water measured as $MFI_{0.1}$ as a function of coagulant dose (mg Fe(III)/L) and Gt at pH 8

For all coagulant doses, $MFI_{0.1}$ was lower at prolonged flocculation time. This effect was particularly pronounced at lower doses, where agglomeration of colloidal iron species into larger aggregates was promoted by increase in flocculation time. At high coagulant dose, volume fraction (φ_i) is larger and as a consequence, the rate of orthokinetic aggregation is higher for a given G-value.

Permeate quality was analyzed in terms of DOC, UV_{254} and SUVA as a function of flocculation time at coagulant dose 0, 5 and 15 mg Fe(III)/L (Table 4-3 and Figure 4-2). Pre-filtered canal water contained approximately 16 mg C/L and a SUVA value of 2.9 L/mg.m. When no coagulant was dosed, DOC was not removed by filtration through 0.1 μm. For coagulated samples, DOC removal was influenced by coagulant dose rather than flocculation time. This observation is in concert with findings of Meyn et al. (2008) for DOC removal with $FeCl_3$ for different flocculation configurations and time.

Table 4-3 Permeate quality of coagulated Delft canal water as a function of flocculation Gt at coagulant dose of 0, 5 and 15 mg Fe(III)/L and pH 8

Exp.	0 mg Fe(III)/L			5 mg Fe(III)/L			15 mg Fe(III)/L		
	DOC mg/L	UV_{254} 1/cm	SUVA L/mg.m	DOC mg/L	UV_{254} 1/cm	SUVA L/mg.m	DOC mg/L	UV_{254} 1/cm	SUVA L/mg.m
a	-	-	-	13.9	0.37	2.7	13.5	0.32	2.4
b	16.3	0.43	2.7	14.4	0.38	2.6	13.1	0.31	2.3
c	15.2	0.44	2.9	14.4	0.37	2.6	13.3	0.31	2.3
d	16.2	0.44	2.7	14.3	0.37	2.6	13.8	0.31	2.3
e	16.5	0.44	2.6	14.5	0.37	2.6	13.3	0.30	2.3
f	16.1	0.44	2.7	14.3	0.38	2.7	13.8	0.31	2.3

At coagulant dose of 5 and 15 mg Fe(III)/L, average DOC removal increased to 11% ± 1% and 16% ± 2% respectively. At pH 8 and coagulant concentration ≥ 5 mg Fe(III)/L, sweep floc is expected to be the dominant coagulation mechanism and DOC is removed by adsorption to and/or enmeshment in iron hydroxide precipitates (Dennett et al., 1996). It may be concluded that adsorption capacity of iron hydroxide itself dictates DOC adsorption, since enhancing agglomeration through increased flocculation time did not affect removal rates. Corresponding residual iron in permeate samples was in the range of 0.03 mg Fe(III)/L, indicating that most iron was precipitated and retained by the membrane.

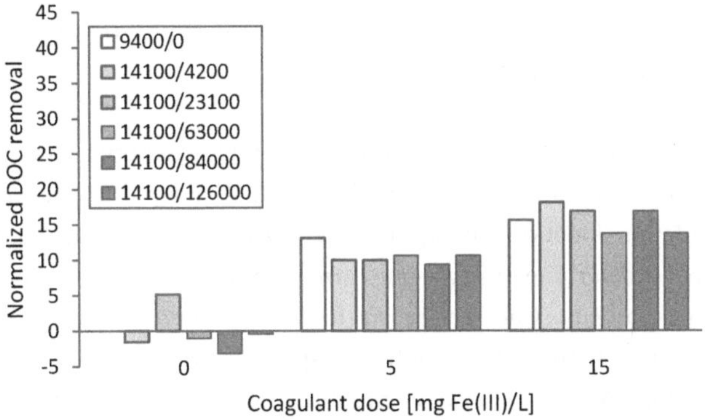

Figure 4-2 Effect of prolonged flocculation time on DOC removal as a function of coagulant dose

4.4.1.2 Effect of pH

Coagulation with $FeCl_3$ is known to be more effective at low pH in terms of DOC removal (Dempsey et al., 1984; Dennett et al., 1996; Jarvis et al., 2003). It is of interest to know whether pH improves the fouling potential as well. The influence of pH on fouling potential was assessed at fixed mixing and flocculation and varying coagulant dose (0, 1 and 5 mg Fe(III)/L). Feed water pH was adjusted prior to coagulation in order to obtain a final pH of 6, 7, and 8. Filtration was performed at 21°C and an applied pressure of 1 bar. Details of mixing and flocculation are given in Table 4-4.

Table 4-4 Mixing and flocculation details for experiments at different pH values

Rapid Mixing			Flocculation			Gt
G [s^{-1}]	t [s]	Gt	G [s^{-1}]	t [min]	Gt	
470	30	14100	0	0	0	14100/0
470	30	14100	35	30	63000	14100/63000
470	30	14100	35	60	126000	14100/126000

$MFI_{0.1}$ values are presented in Figure 4-3 as a function of pH for different flocculation time and coagulant dose. When no coagulant was applied, pH did not affect the fouling potential of Delft canal water as measured by $MFI_{0.1}$.

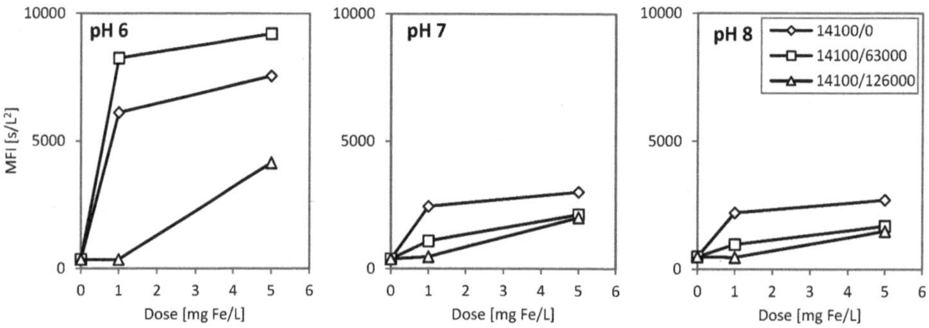

Figure 4-3 $MFI_{0.1}$ as a function of pH, at constant rapid mixing, different flocculation time and coagulant dose of 0, 1 and 5 mg Fe(III)/L and flocculation times

For coagulated solutions, fouling potential was consistently lower at pH 7 and pH 8 as compared with pH 6. Difference in fouling potential between pH 7 and pH 8 was marginal. This was attributed to flocculation kinetics. Zhang et al. (2004) demonstrated that flocculation rate is retarded at lower pH. Lower rate of aggregate formation may result in the formation of smaller aggregates with low porosity. Cake/gel layers formed from such aggregates will have high specific resistance and high MFI. Another aspect is the role of organic matter in coagulation/flocculation. At low pH, better organic matter removal might retard flocculation kinetics; humics are competing with the rate of hydrolysis or precipitate formation. Jarvis et al. (2006) observed that aggregate size decreased when aggregates were formed in the presence of organic matter. Dennett et al. (1996) reported that aggregate formation was slower at low pH resulting in the formation of small flocs that showed poor settling characteristics but substantial filterability. Guigui et al. (2002) reported that flocs formed at low pH and dose, were denser and less porous than those formed at higher pH and dose.

At high flocculation time, lower $MFI_{0.1}$ values were observed. For pH 7 and pH 8, this effect was particularly pronounced at 1 mg Fe(III)/L where lower iron concentration is available for precipitation. At pH 6, when flocculation time was increased from 0 to 30 minutes, higher fouling potential was observed. Further increase in flocculation time to 60 minutes resulted in substantially lower fouling potential as measured by $MFI_{0.1}$.

At pH 6, complexation of organic matter by positively charged iron species results in the formation of soluble and insoluble Fe-OM complexes and iron hydroxide precipitation is retarded. Flocculation for 30 minutes allows simultaneous precipitation of iron hydroxide and adsorption/enmeshment of soluble Fe-OM complexes and other colloidal species to Fe-OM aggregates or precipitated iron hydroxide thereby increase C_b. As a consequence, higher $MFI_{0.1}$ values are observed. Further aggregation for up to 60 minutes results in larger Fe-OM aggregates and lower fouling potential. These observations further suggest that slower flocculation kinetics may have been compensated by prolonged flocculation time, resulting in the formation of larger aggregates that impart less resistance to filtration.

4.4.2 North Sea water

Fouling potential of North Sea water without coagulation was approximately 120 s/L^2. This value is very low for seawater from an open intake. When coagulant was applied, MFI was found to be higher for all but one set of process conditions than that of seawater in the absence of coagulant. Fouling potential of North Sea water coagulated at pH values of 6 and 7.2, and coagulant dose of 5 mg Fe(III)/L was evaluated. Rapid mixing was performed at an intensity of 1,100 s^{-1} for 20 seconds (Gt 22,000). Flocculation time varied from 0 to 45 minutes at an intensity of 40 s^{-1}; Gt 0 and Gt 108,000 respectively. Results are presented in Table 4-5.

Table 4-5 MFI_{300kDa} as a function of pH and flocculation Gt for coagulation with 5 mg Fe(III)/L

	MFI [s/L^2]	
	pH 6	pH 7.2
Uncoagulated seawater	120	120
5 mg Fe(III)/L, Gt 22000/0	280	290
5 mg Fe(III)/L, Gt 22000/108000	160	110

MFI was consistently higher when no flocculation was applied. Increasing flocculation time from 0 minutes to 45 minutes resulted in lower MFI values for both pH 6 and 7.2. This is attributed to the formation of larger aggregates when flocculation is applied for longer times. Aggregate size is inversely proportional to specific cake resistance 'α' and MFI. When only rapid mixing was applied, pH did not affect MFI values. For flocculation time of 45 minutes (Gt 22000/108000), MFI was lower at higher pH.

An attempt was made to restore permeability by scraping off the cake layer with a soft sponge and rinsing the membrane filters in Milli-Q water. Figure 4-4 shows results for initial and final membrane resistance (R_m) for different treatment conditions. The difference in the two values is considered to be permeability loss and might be attributed to pore blocking. Highest permeability loss was observed for direct seawater filtration. Coagulation substantially reduced the loss in permeability. This observation was attributed to reduction in pore blocking due to the creation of larger aggregates by coagulation. Flocculation applied at pH 6 for 45 minutes, gave better results. At pH 7.2 the opposite effect was observed. From these observations it cannot be generally concluded that longer flocculation time reduces permeability loss.

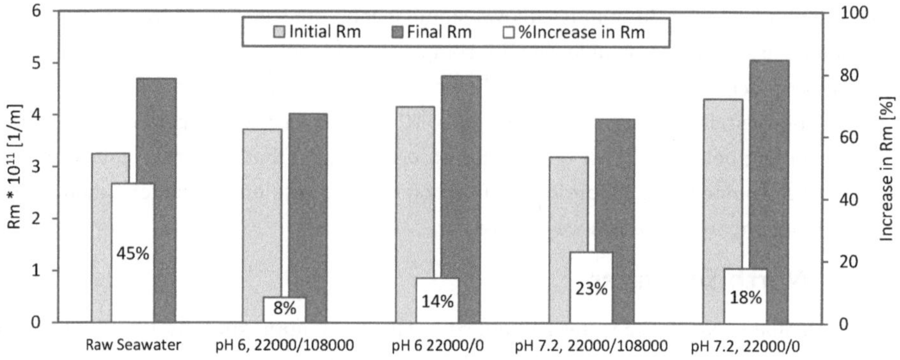

Figure 4-4 Increase in membrane resistance (R_m) as a result of pore blocking (after removal of cake/gel layer) for different operational conditions

Scraping off the cake/gel layer to get an impression of permeability loss due to pore blocking is informative. However in the next chapters experiments comprising successive filtration and backwash cycles is preferred, since these tests give direct insight in the development of non-backwashable fouling.

4.4.3 Synthetic seawater spiked with AOM

As North Sea water did not exhibit high fouling potential in terms of MFI, experiments were performed with solutions of AOM produced by the marine diatom *Chaetoceros affinis* in synthetic seawater. Harvested AOM stock solution was diluted with synthetic seawater to create feed solutions at a concentration of 0.7 mg C/L as biopolymers. Results of LC-OCD analysis of harvested AOM stock solution and feed solution are given in Table 4-6. The theoretical cut-off of the method is 2 µm, which means that particles larger than 2 µm are not taken into account in these measurements. However, it cannot be excluded that part of the sticky AOM (e.g., TEP) would also be rejected, even at particle size < 2 µm.

Table 4-6 DOC fractionation for AOM stock solution and feed solution diluted to 0.7 mg C/L as biopolymers

AOM Stock Solution						
DOC	Biopolymers (>> 20,000 Da)	Humics (~ 1000 Da)	SUVA	Building Blocks (300 - 500 Da)	LMW Neutrals (< 350 Da)	LMW Acids (< 350 Da)
mg C/L	mg C/L	mg C/L	L/mg.m	mg C/L	mg C/L	mg C/L
5.72	2.51	1.87	0.94	0.34	0.96	0.40
Feed Solution (AOM in Synthetic Seawater)						
1.16	0.66	0.45	0.77	0.14	0.28	0.18

Villacorte (2014) attributed the presence of humic substances and low molecular weight compounds to the composition of the medium in which the diatoms are grown.

4.4.3.1 Effect of dose and pH

Experiments were designed to investigate the effect of pH, dose and flocculation G and Gt on AOM fouling potential. All other parameters i.e., temperature, rapid mixing G and Gt, salinity, and AOM concentration were kept constant.

The effect of coagulant dose and pH on AOM coagulation was evaluated with MFI_{100kDa}. Coagulation was performed at 0.5 and 2 mg Fe(III)/L, at two different pH values of 6 and 8. Rapid mixing was done at 1,100 s^{-1} for 20 seconds (RM Gt 22,000), followed by slow mixing at 45 s^{-1} for 30 minutes (SM Gt 81,000). Results are shown in Figure 4-5.

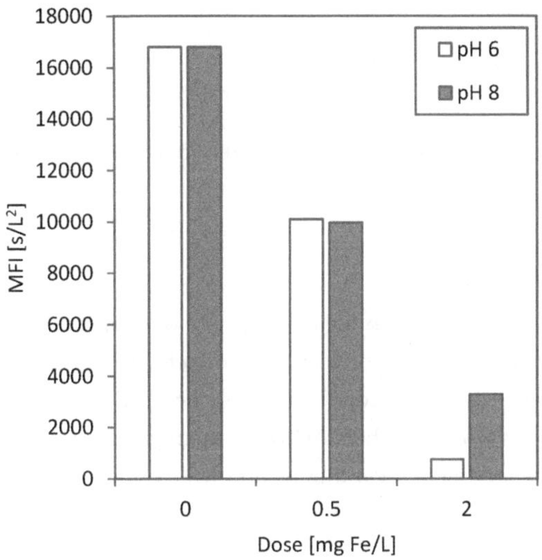

Figure 4-5 MFI of coagulated AOM as a function of coagulant dose and pH at Gt 22000/81000

MFI_{100kDa} was high for AOM solutions in synthetic seawater, in comparison with MFI values obtained for Delft Canal Water and North Sea water. It must be noted that in MFI measurements, smaller pore size (or molecular weight cut-off) results in higher MFI values as more of the smaller particles can be retained.

The high MFI of AOM solutions may be attributed to the formation of cake/gel layers with low permeability. Addition of coagulant resulted in lower MFI values; the higher the dose, the lower the MFI_{100kDa} for both pH values of 6 and 8. The effect of pH at coagulant dose of 0.5 mg Fe(III)/L was negligible. However, at 2 mg Fe(III)/L, lower pH resulted in lower fouling potential. Filtration curves of dt/dV versus V as a function of dose and pH are presented in Figure 4-6.

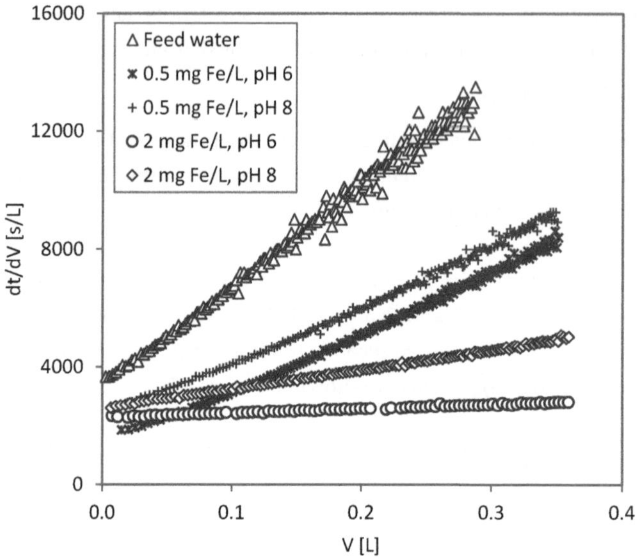

Figure 4-6 Filtration curves (dt/dV versus V) for different coagulant conditions, Gt 22000/81000, feed biopolymer concentration 0.7 mg C/L

4.4.3.2 Effect of flocculation

Longer flocculation times may result in the enlargement of coagulated particles. Larger particles are expected to impart lower resistance to filtration (refer Chapter 3, section 3.2.1 and 3.2.2), firstly, due to a size effect (d_p) and secondly, because larger flocs have higher porosity (ε). It is therefore of interest to verify whether AOM-Fe aggregates grow in size with prolonged flocculation times. The experimental conditions and MFI values are presented in Table 4-7 and Figure 4-7 respectively.

Table 4-7 Effect of flocculation time on fouling potential of AOM at pH 8

Rapid mixing			Flocculation			Gt
G [s^{-1}]	t [s]	Gt	G [s^{-1}]	t [min]	t [min]	
1100	20	22000	0	0	0	22000/0
1100	20	22000	45	30	81000	22000/81000

At low coagulant dose of 0.5 mg Fe(III)/L, the effect of flocculation time is negligible. At this dose, flocculation time of 30 minutes did not improve MFI, indicating that the iron concentration is too low to destabilize and/or enmesh small particles adequately. Alternatively destabilization might have occurred but further aggregation did not result in an improvement of MFI.

At 2 mg Fe(III)/L, flocculation time of 30 minutes, resulted in an improvement in MFI of approximately 15%. It may be that some aggregation and growth occurs, resulting in a marginal improvement in MFI. However, it may also be possible that further increase in size of flocculated AOM did not improve the permeability of their deposited cake. Particle size analysis could provide more insight on the effect of flocculation by prolonged flocculation on the aggregation of coagulated AOM particles.

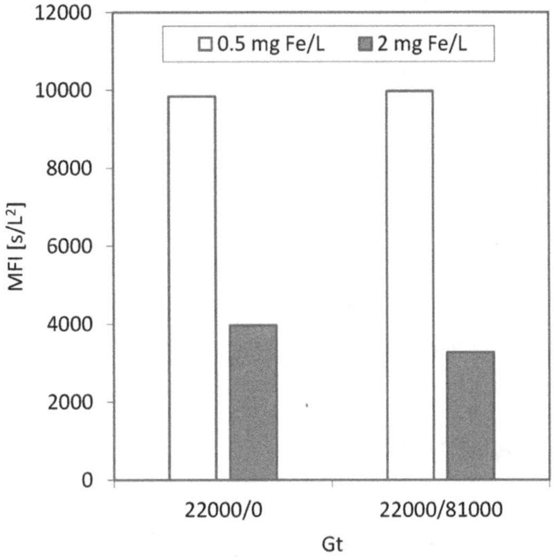

Figure 4-7 Effect of flocculation time on fouling potential of coagulated AOM at pH 8

As no apparent effect of flocculation time on permeability of coagulated AOM was observed, the effect of flocculation Gt was investigated. Flocculation G-values of 45 s^{-1} and 200 s^{-1} were chosen, and flocculation time was varied to achieve constant flocculation Gt of 24,000, 84,000 and 180,000. Details are provided in Table 4-8. Coagulation was performed at pH 8 with 2 mg Fe(III)/L, rapid mixing intensity of 1100 s^{-1} and rapid mixing time of 20 seconds.

Table 4-8 Experimental details to determine effect of flocculation G and Gt on AOM fouling potential

Flocculation G [s^{-1}]	Flocculation t [min]	Flocculation Gt	Flocculation G [s^{-1}]	Flocculation t [min]	Flocculation Gt
45	0	0	200	0	0
45	9	24000	200	2	24000
45	30	84000	200	7	84000
45	67	180000	200	15	180000

Results indicate that flocculation up to 115 minutes at G-value of 45 s^{-1} does not have a pronounced effect (Figure 4-8). At a G-value of 200 s^{-1} after 7 minutes flocculation, a substantially lower MFI is achieved. However, after prolonged flocculation a higher MFI was observed. This might be attributed to a grinding effect due to the relatively high G-value for flocculation.

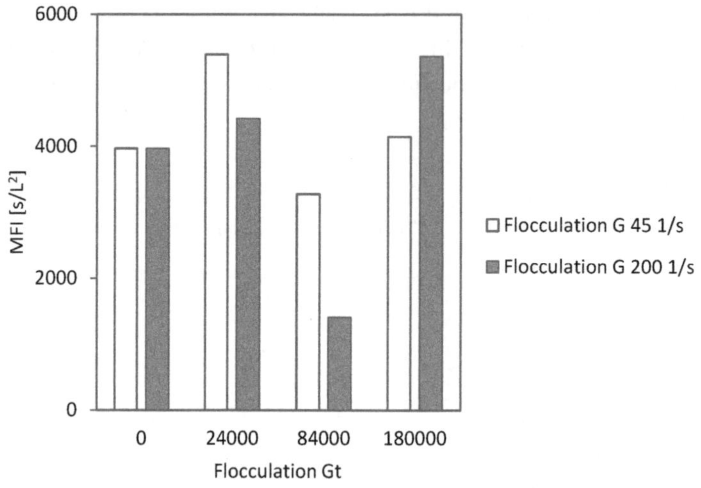

Figure 4-8 MFI as a function of flocculation Gt at pH 8, 2 mg Fe(III)/L and AOM concentration 0.7 mg C/L as biopolymers

4.4.4 G and Gt in MF/UF systems

The fate of aggregates in the pipe network prior to MF/UF membranes and inside the elements is of interest in judging the development of fouling potential of the feed water, since flow regimes in the pipes and modules might favour growth and/or breakage of aggregates.

Studies on turbulent pipe flow for particle destabilization and aggregation show that for Reynolds number between 8,000 and 16,000, reaction rate for particle aggregation increased (Klute, 1977). Beyond 16,000 however, reaction rate decreased with increasing Reynolds number. This effect may be attributed to a reduced collision efficiency of the primary particles and/or disruption of (micro) flocs if the turbulence intensity in the pipe reactor exceeds a certain critical value. Reports on flocculation experiments in pipes of various diameters (8 to 600 mm) show that under steady-state conditions, a smaller floc size is observed at higher flow velocity (Grohmann, 1985).

This section highlights the fate of flocculated particles throughout the treatment process. Regardless of the conditions in which flocculation is achieved, exposure of flocculated aggregates to G and Gt values in pumps, pipes, valves and even within MF/UF capillaries may break the aggregates or affect their structure. To elaborate this concept, calculations were made for G and Gt values within the pipe network and hollow fibres of an arbitrary treatment plant considering the scheme presented in Figure 4-9. The MF/UF system is comprised of three skids in parallel which operate in dead end mode; each skid accommodates 12 elements (6x2). Considering a membrane area of 35 m^2 per element, the total filtration area of each skid is 420 m^2. Therefore, at a permeate flux of 80 L/m^2h, the flow to each skid is 33 m^3/h. With reference to this flow rate, velocities in different pipe sections in the system were calculated (Table 4-9).

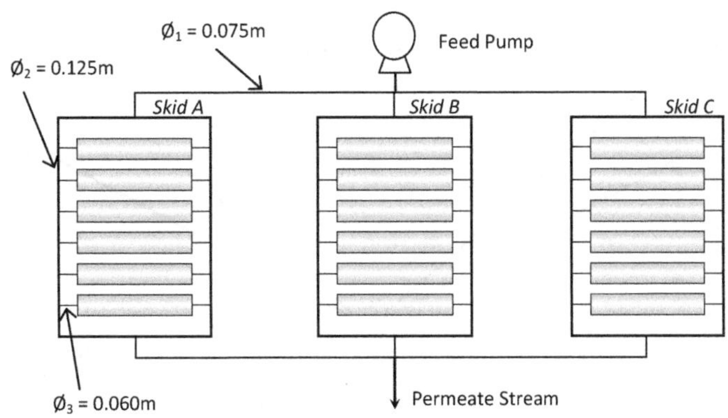

Figure 4-9 Scheme of MF/UF system in an arbitrary full scale plant

Calculations revealed turbulent flow regimes for all pipe diameters. G-values in the pipes (refer Eq. 4-6) were between 700 s^{-1} to 9600 s^{-1}, depending on pipe diameter. Klute and Amirtharajah (1991) proposed G-values in the range of 200-500 s^{-1} for particle aggregation to form microflocs and 50-100 s^{-1} for particle aggregation into macroflocs. Overall retention time up to the first module in Skid A is about 15 s, resulting in a total Gt of approximately 71,000.

Table 4-9 Velocity and Reynolds number in the pipe network of a treatment plant

Pipe	Diameter (Ø)	Velocity	Reynolds No.	G
	[m]	[m/s]	[]	[s^{-1}]
1	0.075	2	155000	4000
2	0.125	0.75	93000	700
3	0.060	3	193400	9600

For a single hollow fibre fed from both ends, flow rate, velocity and G values were calculated at a flux of 80 L/m²h. The hollow fibre was divided into 20 arbitrary sections of equal length (0.075 m), for better elaboration of G and Gt development along the fibre length (Figure 4-10).

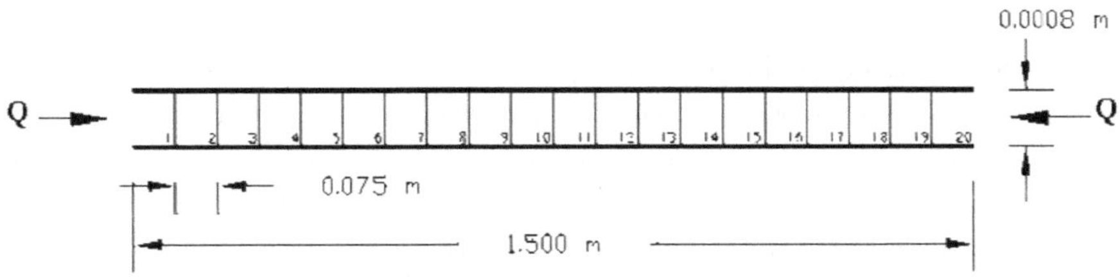

Figure 4-10 Longitudinal section of an MF/UF capillary

For each arbitrary section of the fibre, the mean flow and head loss were calculated. Reduction in flow is based on the assumption that permeation is constant in all sections. Based on head loss, G was calculated for each section. Results are illustrated in Figure 4-11.

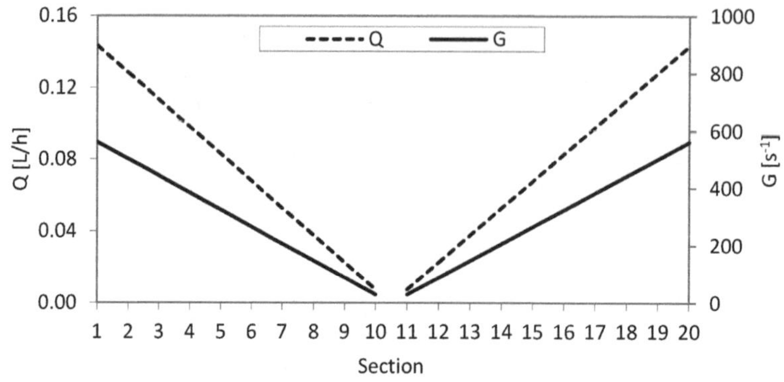

Figure 4-11 Change in average flow and G-value in arbitrary sections of equal length along an MF/UF capillary

Calculated average velocities were around 0.08 m/s at the fibre entrance and approached zero at the centre of the fibre. Calculations showed that G-value decreased proportionally with decreasing flow along the fibre length; such that G was at its maximum at the fibre entrance at approximately 560 s^{-1}. This value decreased along the fibre, till it reached its minimum at 0 s^{-1}. Residence time in the fibre is about 18 s.

G-values in the first few sections of the fibre could be high enough to break the aggregates. However the aggregates present are formed at much higher G-values, consequently breakage is unlikely to occur and it is unlikely that they can re-form due to the very short residence time in the fibre. Further calculations showed that aggregates entering the fibres can be exposed to a total Gt of about 5,500 by the time they reach fibre mid-section.

Results of experiments at high G-values followed by flocculation at low G-values indicated that the total Gt to which particles/aggregates are exposed - amounting to approximately 77,000 - is sufficient to prevent the occurrence of high fouling potential (MFI).

4.5 Conclusions

Coagulant addition resulted in higher fouling potential for almost all tested conditions, even at high coagulant dose and prolonged flocculation time where large aggregates are likely formed, creating cakes with low specific resistance. This was due to the very low fouling potential of fresh water and real seawater. However, the absolute values of fouling potential as measured by MFI were low, as the range of MFI results in low predicted pressure increase in MF/UF systems.

In fresh water, flocculation partly compensated the deterioration of fouling potential. This was particularly the case for long flocculation time at low dose.

In synthetic seawater spiked with algal organic matter, coagulation demonstrated a very pronounced improvement of the fouling potential. For coagulant addition and rapid mixing only, the fouling potential was substantially improved. Subsequent flocculation up to 30 minutes demonstrated marginal improvements.

Coagulation at pH 6 resulted in 10 to 20 times higher fouling potentials for fresh water. At pH 7 and 8 this effect was much less pronounced; 2 to 3 times higher fouling potentials. In real seawater no difference between pH 6 and 7 was observed. However in synthetic seawater spiked with AOM - at coagulant dose of 2 mg Fe(III)/L - pH 6 gave better results than pH 8. The observed inconsistency in results might be attributed to the relatively high concentration of humic acids in fresh water as opposed to synthetic seawater spiked with AOM.

Results indicated that absence of flocculation at moderate G-values as applied commonly in conventional systems can be compensated - necessary - by applying higher coagulant dose. Calculations showed very high G-values and short residence time prior to and within MF/UF elements, which are likely sufficient in maintaining/achieving relatively low fouling potential of feed water reaching the membrane surface. The results of this chapter suggest that the main mechanism involved in fouling control in inline coagulation is most likely the improvement of backwashability of the fouling layer.

4.6 References

Alldredge, A. L., Passow, U., & Logan, B. E., 1993. The abundance and significance of a class of large, transparent organic particles in the ocean. Deep-Sea Research I, 40, 1131-1140.

Bache, D.H., Johnson, C., Papavasilopoulos, E., Rasool, E., McGilligan, J.F., 1999. Sweep coagulation: structures mechanisms and practice. Journal of Water Science Research and Technology-AQUA 48 (5), 201-210.

Bache, D.H., Rasool, E., Moffatt, D., McGilligan, F.J., 1999. On the strength and character of alumino-humic flocs. Water Science and Technology 40 (9), 81–88.

Berman, T., Mizrahi, R., Dosoretz, C. G., 2011. Transparent exopolymer particles (TEP): A critical factor in aquatic biofilm initiation and fouling on filtration membranes. Desalination 276(1-3), 184-190.

Bernhardt, H., Hoyer, O., Schell, H., Lusse, B., 1985. Reaction mechanisms involved in the influence of algogenic matter on flocculation. Zeitschrift fur Wasser und Abwasser forschung 18, 18-30.

Bratby, J., 2006. Coagulation and flocculation in water and wastewater treatment. Second Edition. London: IWA Publishing.

Busch, M., Chu, R. & Rosenberg, S., 2010. Novel trends in dual membrane systems for seawater desalination: minimum primary pre-treatment and low environmental impact treatment schemes. IDA Journal, Desalination and Water Reuse 2(1), 56-71.

Camp, T.R, Stein, P.C., 1943. Velocity gradients and internal work in fluid motion. Journal of the Boston Society of Civil Engineers, October.

Camp, T.R., 1955. Flocculation and flocculation basins. Transactions, American Society of Civil Engineers 120, 1-16.

Dempsey, B.A., Ganho, R.M., O'Melia, C.R., 1984. The coagulation of humic substances by means of aluminium salts. Journal of the American Water Works Association 76(4), 141–150.

Dennett, K.E., Amirtharajah, A., Moran, T.F., Gould, J.P., 1996. Coagulation: its effect on organic matter. Journal of the American Water Works Association 84(4), 129–142.

Edzwald, J.K., Haarhoff, J., 2011. Seawater pretreatment for reverse osmosis: Chemistry, contaminants, and coagulation. Water Research 45, 5428-5440.

EPA, USEPA, 1999. Enhanced Coagulation and Enhanced Precipitative Softening Guidance Manual. Disinfectants and Disinfection Byproducts Rule (DBPR), 237.

Gabelich, C., Ishida, K.P., Gerringer, F.W., Evangelista, R., Kalyan, M., Suffet, I.H., 2006. Control of residual aluminium from conventional pretreatment to improve reverse osmosis. Desalination 190, 147-160.

Gregor, J.E., Nokes, C.J., Fenton, E., 1997. Optimising natural organic matter removal from low turbidity waters by controlled pH adjustment of aluminium coagulation. Water Research 31 (12), 2949–2958.

Grohmann, A., 1985. Flocculation in pipes: design and operation. In: Grohmann, A., Hahn, H.H., Klute. R. (Eds.) Chemical water and wastewater treatment. Stuttgart: Gustav Fischer Verlag.

Guigui, C., Rouch, J.C., Durand-Bourlier, L., Bonnelye, V., Aptel, P., 2002. Impact of coagulation conditions on the inline coagulation/UF process for drinking water production. Desalination 147, 95-100.

Halpern, D.F., McArdle, J. & Antrim, B., 2005. UF pre-treatment for SWRO: pilot studies. Desalination 182, 323-332.

Huber, S.A., Balz, A., Abert, M., Pronk, W., 2011. Characterisation of aquatic humic and non-humic matter with size-exclusion chromatography - organic carbon detection - organic nitrogen detection (LC-OCD-OND). Water Research 45, 879-885.

Huber, S.A., Frimmel, F.H., 1994. Characterization and quantification of marine dissolved organic carbon with a direct chromatographic method. Environmental Science and Technology 28, 1194-1197.

Jarvis, P., Jefferson, B., Parsons, S.A., 2006. Floc structural characteristics using conventional coagulation for a high doc, low alkalinity surface water source. Water Research 40, 2727-2737.

Klute, R., 1977. Adsorption von polymeren and silicaoberflachen bei unterschiedlichen stromungsbedingungen. Doctoral dissertation, University Karlsruhe, Germany.

Klute, R., Amirtharajah, A., 1991. Particle destabilization and flocculation reactions in turbulent pipe flow. In: Amirtharajah, A., Clark, M.M., Trussel, R.R. (Eds.) Mixing in coagulation and flocculation. Colorado, AWWA Research Foundation.

Lattemann, S., 2010. Development of an environmental impact assessment and decision support system for seawater desalination plants. Doctoral dissertation, UNESCO-IHE/TUDelft, Delft.

Leiknes, T.O., 2009. The effect of coupling coagulation and flocculation with membrane filtration in water treatment: A review. Journal of Environmental Sciences 21, 8-12.

Lyman, J., Fleming, R.H., 1940. Composition of seawater. Journal of Marine Research 3, 134-146.

Meyn, T., Bahn, A., Leiknes, T., 2008. Significance of flocculation for NOM removal by coagulation - ceramic membrane microfiltration. Water Science and Technology: Water Supply 8(6), 691-700.

Monk, R., Trussell, R., 1991. Design of mixers in water treatment plants: Rapid mixing and flocculation. In: Amirtharajah, A., Clark, M.M., Trussel, R., (Eds.) Mixing in coagulation and flocculation. Denver: AWWA Research Foundation.

O' Melia, C.R., Stumm, W., 1967. Aggregation of silica dispersion by iron (III). Journal of Colloid and Interface Science 23 (3), 437-447.

Passow, U., Alldredge, A.L., 1994. Distribution, size, and bacterial colonization of transparent exopolymer particles (TEP) in the ocean. Marine Ecology Progress Series 113, 185-198.

Passow, U., Alldredge, A.L., 1995. A dye-binding assay for the spectrophotometric measurement of transparent exopolymer particles (TEP). Limnology and Oceanography 40, 1326-1335.

Pearce, G.K., 2011. The role of coagulation in membrane pre-treatment for seawater desalination. In International Desalination Association World Congress, Perth, Australia.

Pivokonsky, M., Kloucek, O., Pivokonska, L., 2006. Evaluation of the production, composition and aluminium and iron complexation of algogenic organic matter. Water Research 40, 3045-3052.

Shen, Y.H., Dempsey, B.A., 1998. Fractions of aluminium coagulants as a function of pH. Environmental Technology 19, 845–850.

Stumm, W., Morgan, J.J., 1962. Chemical aspects of coagulation. Journal of the American Water Works Association 64 (8), 971-993.

Ventresque, C., Gisclon, V., Bablon, G., Chagneau, G., 2000. An outstanding feat of modern technology: the Mery-sur-Oise nanofiltration treatment plant (340,000 m3/d). Desalination 131, 1-16.

Villacorte, L.O., 2014. Algal blooms and membrane based desalination technology. Doctoral dissertation, UNESCO-IHE/TUDelft, Delft.

Villacorte, L.O., Schurer, R., Kennedy, M., Amy, G., Schippers, J.C., 2010. The fate of transparent exopolymer particles in integrated membrane systems: a pilot plant study in Zeeland, The Netherlands. Desalination and Water Treatment 13, 109-119.

von Smoluchowski. M., 1917. Versuch eine mathematischen Theorie der Koagulationskinetik kolloidaler Losungen. Zeitschrift fur Physikalische Chemie 92, 129-168.

Wolf, P.H., Siverns, S. & Monti, S., 2005. UF membranes for RO desalination pre-treatment. Desalination 182, 293-300.

Zhang, P., Hahn, H.H., Hoffmann, E., 2004. Study on flocculation kinetics of silica particle suspensions. In: Hahn, H.H., Hoffmann, E., Odegaard, H. (Eds.) Chemical water and wastewater treatment. First Edition. Florida: IWA Publishing.

Zhou, J., Mopper, K., Passow, U., 1998. The role of surface-active carbohydrates in the formation of transparent exopolymer particles by bubble adsorption of seawater. Limnology and Oceanography 43(8), 1860-1871.

5

FOULING POTENTIAL AND REMOVAL OF ALGAL ORGANIC MATTER IN SEAWATER ULTRAFILTRATION

This chapter investigated the effect of coagulation on fouling potential and removal of algal organic matter (AOM) in seawater ultrafiltration (UF) systems. AOM harvested from a strain of bloom forming marine diatom, *Chaetoceros affinis*, was coagulated with ferric chloride under different coagulation modes and conditions. AOM (as biopolymers) had a high fouling potential as measured by MFI-UF, which strongly depended on filtration flux. Moreover, the developed cake/gel layer on the membrane was fairly compressible during filtration; manifested as higher fouling potential at higher filtration flux and non-linear development of pressure in filtration tests. Coagulation substantially reduced fouling potential and compressibility of the AOM cake/gel layer. The impact of coagulation was particularly significant at coagulant dose of 1 mg Fe(III)/L and higher. Coagulation also substantially reduced the flux-dependency of AOM fouling potential, resulting in linear development of pressure in filtration tests at constant flux. This was attributed to adsorption of biopolymers on precipitated iron hydroxide and formation of Fe-biopolymer aggregates, such that the fouling characteristics of iron hydroxide precipitates prevailed and AOM fouling characteristics diminished. At low coagulant dose, inline coagulation/UF was more effective in removing AOM than the other two modes tested. At high coagulant dose where sweep floc conditions prevailed, AOM removal was considerably higher and controlled by coagulant dose rather than coagulation mode.

This chapter is a modified version of:

Tabatabai, S.A.A., Schippers, J.C., Kennedy, M.D., 2014. Effect of coagulation on fouling potential and removal of algal organic matter in ultrafiltration pretreatment to seawater reverse osmosis. Water Research, *In Press*.

5.1 Introduction

Micro- and ultrafiltration (MF/UF) are rapidly gaining ground as alternatives to conventional processes such as dual media filtration in seawater reverse osmosis (SWRO) pretreatment. Although MF/UF offer several advantages over conventional pretreatment systems, such as lower footprint, lower overall chemical consumption, better permeate quality, etc. (Wilf and Schierach, 2001; Pearce, 2007), membrane fouling remains a major obstacle to high production capacity of these systems (Pearce, 2009). In seawater applications, UF fouling is mainly associated with periods of algal bloom during which UF operation is characterized by rapid permeability decline, poor backwashability and high chemical cleaning frequency of the UF membranes (Schurer et al., 2013). Villacorte (2014) attributed the deterioration in UF hydraulic performance to algal cells and their associated algal organic matter (AOM) prevalent during algal bloom conditions.

AOM produced by common bloom-forming algae largely comprises high molecular weight biopolymers, i.e., polysaccharides and proteins (Fogg, 1983; Myklestad, 1995). A significant component of AOM is transparent exopolymer particles (TEP) (Mopper et al., 1995; Villacorte et al., 2013) which are highly sticky acid polysaccharides and glycoproteins (Passow and Alldredge, 1994). AOM can therefore adhere strongly to the surface or within the pores of UF membranes providing substantial resistance to permeate flow during filtration and reducing the efficiency of hydraulic cleaning in restoring permeability (Villacorte et al., 2010; Qu et al., 2012 a & b; Schurer et al., 2013). Moreover, AOM that pass through UF membranes can initiate and exacerbate particulate/organic fouling and biofouling in downstream reverse osmosis membranes (Berman and Holenberg, 2005; Berman et al., 2011).

To control fouling in UF systems, coagulation is commonly applied using metal salts, e.g. aluminium and iron (Busch et al., 2010). Coagulation of feed water can reduce transmembrane pressure (TMP) development and the extent of non-backwashable fouling in UF systems (Guigui et al., 2002; Lee et al., 2000; Choi and Dempsey, 2004). Effective coagulation conditions result in reduced pore blocking, higher permeability of the cake/gel layer formed on the membrane surface and reduced strength of adhesion of particles to the membrane surface, by altering particle properties such as size, thereby improving hydraulic performance of the UF membranes. Coagulation can also enhance permeate quality by enhancing dissolved organic matter removal (Guigui et al., 2002).

Coagulation of organic matter is complex and the mechanisms and reaction pathways are highly pH-dependant (Rebhun and Lurie, 1993; Jarvis et al., 2006). Proposed mechanisms for organic matter coagulation are chemical precipitation by complexation with soluble metal species for pH < 6 and adsorption to and/or enmeshment in metal hydroxide precipitates for pH > 6 (Dempsey et al., 1984; Gray, 1988; Dennett et al., 1996). In seawater, where dissolved metal speciation is affected by the high ionic strength, optimum pH values for organic matter complexation may be higher than mentioned above (Duan et al., 2002; Edzwald and Haarhoff, 2011).

Most studies on coagulation in UF systems address fouling due to natural organic matter (NOM) by measuring surrogate parameters such as dissolved organic carbon (DOC) and UV absorbance at 254 nm (Choi and Dempsey, 2004; Cho et al., 2006; Howe and Clark, 2006). However, Henderson et al. (2008) demonstrated that AOM is of a very different character to NOM and therefore existing knowledge on NOM coagulation may not be adequate to explain the effect of coagulation on AOM fouling potential and removal in UF systems. A recently developed surrogate for quantifying AOM concentrations in freshwater and seawater is biopolymers concentration measured by liquid chromatography-organic carbon detection (LC-OCD) (Huber and Frimmel, 1994; Huber et al., 2011). In this technique biopolymers are defined as high molecular weight compounds that do not absorb UV_{254} and contain nitrogen to varying degrees. These characteristics point to the presence of polysaccharides and proteins, the main components of AOM. Recent studies have shown remarkable correlations between biopolymers concentration as measured by LC-OCD in different surface waters with reversible and irreversible UF membrane fouling (Tian et al., 2013; Kimura et al., 2014).

This work investigated the effect of coagulation on fouling potential and removal of AOM in UF systems. Specifically, AOM obtained from a marine diatom - *Chaetoceros affinis* - was coagulated under different conditions of dose, mixing and flocculation and the fouling behaviour and removal rates in a UF system were studied. The objectives were:

1. To determine the effect of different coagulation conditions, simulating inline coagulation, on fouling potential and pressure development in UF tests.
2. To determine the effect of coagulation on the removal rates of AOM, measured with LC-OCD, in three modes, namely,
   - Coagulation/flocculation/sedimentation;
   - Coagulation/flocculation/sedimentation/filtration (0.45 μm);
   - Inline coagulation/UF (150 kDa).

For a sound understanding, the behaviour of different waters in MF/UF should be observed from two angles, namely, fouling potential as measured by e.g., the Modified Fouling Index-Ultrafiltration (MFI-UF) and backwashability. This study addressed the effect of coagulation on fouling potential and fouling mechanisms of AOM.

5.2 Theoretical background

Hydraulic performance of UF membranes operated in constant flux mode is assessed based on pressure increase in each cycle, recovery of pressure after hydraulic cleaning and recovery of pressure by chemical cleaning (CEB, CIP) after multiple cycles. Pressure increase in a filtration cycle is due to fouling which is caused by the retention of colloids, particles and to some extent large macromolecules on the membrane surface. Fouling mechanisms in UF systems are identified mainly as (in)compressible cake/gel layer formation and pore blocking (Ye et al., 2005; Jermann et al., 2008).

5.2.1 Cake/gel filtration and fouling potential

Flow through a membrane can be defined based on Darcy's law for flow through porous media,

$$J = \frac{dV}{Adt} = \frac{\Delta P}{\eta(R_m)} \qquad \text{(Eq. 5-1)}$$

Where,
J is filtration flux (m/s)
ΔP is applied pressure (Pa)
A is filtration area (m^2)
η is dynamic viscosity (Pa.s)
R_m is clean membrane resistance (1/m).

In the absence of fouling matter, flow through the membrane is proportional to applied pressure and the permeability of the medium. When fouling matter is present in the water, accumulation on the membrane surface and/or within the pores results in an increase in resistance to filtration. The resistance in series model is commonly used to describe pressure development in UF membranes,

$$J = \frac{\Delta P}{\eta(R_m+R_b+R_c)} \qquad \text{(Eq. 5-2)}$$

Where,
R_b is resistance imparted to flow due to pore blocking (1/m)
R_c is resistance due to cake/gel formation (1/m).

Ruth et al. (1933) defined cake/gel resistance as,

$$R_c = \frac{V}{A}.I \qquad \text{(Eq. 5-3)}$$

Where,
V is filtered volume (m^3)
I is the fouling index (m^{-2}).

For constant flux filtration, assuming R_b does not play a role after a short period from the start of filtration, substituting Eq. 5-3 in Eq. 5-2 and rearranging gives,

$$\Delta P = \eta R_m J + \eta.I.J^2.t \qquad \text{(Eq. 5-4)}$$

Modified Fouling Index (MFI) was originally defined as an index of the fouling potential of RO feed water by Schippers and Verdouw (1980) for filtration at constant pressure using membranes with 0.45 μm pore size as,

$$\mathrm{MFI} = \frac{\eta I}{2\Delta P A^2} \qquad \text{(Eq. 5-5)}$$

Where,
ΔP is reference pressure value of 2 bar
η is reference viscosity at 20 °C (0.001003 Pa.s)
A is reference filtration area (13.8 x 10^{-4} m^2).

Further work by Boerlage et al. (2004) and Salinas Rodriguez (2011) resulted in the development of MFI-UF for filtration at constant flux, using ultrafiltration membranes. In constant flux mode, I, is determined from the slope of the linear region in a plot of ΔP versus time and is characteristic of cake/gel filtration mechanism.

The fouling index, I, is the product of particle concentration in feed water and the average specific resistance 'α' of the cake/gel layer formed on the membrane surface. The specific cake/gel resistance is a function of particle properties such as size and density, and cake/gel properties such as porosity and is defined, for spherical particles, by the Carman-Kozeny relation as,

$$\alpha = \frac{180(1-\varepsilon)}{\rho d_p^2 \Phi^3 \varepsilon^3} \qquad \text{(Eq. 5-6)}$$

Where,
ε is porosity of the cake/gel layer (-)
ρ is particle density (kg/m^3)
d_p is particle size (m)
Φ is particle sphericity.

MFI-UF and I represent bulk properties of particles in feed water and therefore reflect variations in water quality characteristics. As such, these parameters can be used as surrogate parameters to assess the fouling potential of certain compounds in UF systems. In this study MFI-UF measured at constant flux through 150 kDa UF membranes, is used as a tool to assess the fouling potential of AOM, in the absence or presence of coagulant.

5.2.2 Prediction of pressure development in UF capillaries

For constant flux operation, pressure increase throughout a filtration cycle is a function of membrane properties (i.e., pore size) and feed water characteristics. Initially pore blocking is expected to occur resulting in a fast increase in transmembrane pressure (TMP). Quite soon, cake/gel filtration will take place showing a much slower pressure increase. In this stage TMP development follows Eq. 5-4. Later on, compression of the cake/gel layer might occur resulting in a faster development of the TMP.

Since MFI-UF is directly proportional to the fouling index, I, measured MFI (or I) values can be plugged in Eq. 5-4 to predict pressure development in a UF system. The main assumption in predicting UF performance using MFI-UF values is that cake/gel filtration is the dominant mechanism. In practice this may not be the case, in particular when compression of the cake/gel occurs and when backwashability is poor (due to pore blocking).

In this study, experimentally obtained MFI-UF values were plugged in Eq. 5-4 to predict pressure increase in a UF system for AOM filtration in the presence and absence of coagulant. Predictions were compared with actual laboratory data of TMP versus time for different conditions.

5.3 Materials and methods

5.3.1 Algae cultivation and AOM production

To simulate seawater bloom conditions, the marine diatom species *Chaetoceros affinis* (CCAP 1010/27) obtained from the Culture Collection of Algae and Protozoa (Oban, Scotland), were cultivated for AOM production. Diatoms of the genus Chaetoceros are known to produce large quantities of extracellular polysaccharides throughout their growth cycle (Watt, 1968; Dam and Drapeau, 1995; Myklestad, 1995), making them a suitable choice for laboratory-scale production of AOM in short term experiments.

Cultures were grown at room temperature in 2 L flasks containing 1 L of sterilised f/2+Si medium (CCAP, 2010) prepared with autoclaved synthetic seawater with a TDS of 35 g/L. Ionic composition of the SSW followed the recipe of Lyman and Fleming (1940). The cultures were set on continuous shaking at moderate speed (~ 100 rpm) to ensure nutrient distribution and to avoid settling of cells. To enhance AOM excretion, cultures were exposed to continuous (Hellebust, 1965) mercury fluorescent light with an average incident photon flux density of 40 $\mu mol\ m^{-2}s^{-1}$.

AOM was harvested from the diatoms during the stationary phase of growth, approximately 14 days after inoculation when production of extracellular polysaccharides is known to be highest (Myklestad, 1974). Cells were allowed to settle for 48 hours and supernatant containing AOM was separated by suction without disturbing the settled algae at the bottom of the flasks. Samples were not filtered to avoid loss of AOM from the solutions. The harvested AOM is assumed to be mostly extracellular algal organic matter, as no cell destruction was attempted. Harvests from different batches were stored in a 20 L glass bottle and fully mixed to obtain one stock AOM solution for characterization and preparation of feed solutions.

5.3.2 AOM characterization

5.3.2.1 LC-OCD

Organic carbon in the AOM stock solution was quantified and fractionated by LC-OCD (DOC-Labor, Germany). In this method, separation is based on size-exclusion chromatography followed by multi-detection of organic carbon, UV absorbance at 254 nm and organic nitrogen. Chromatograms are processed on the basis of area integration using the programme ChromCALC. The method separates NOM according to size/molecular weight and can give both quantitative and qualitative information on up to 10 classes of natural organics. The definitions and size ranges formally assigned to different fractions of the LC-OCD analysis are given by Huber et al. (2011). In the LC-OCD method, DOC is measured in the column bypass after inline filtration through 0.45 μm. However, AOM and its constituents e.g., TEP, can be much larger. Therefore, pre-filtration was bypassed to avoid loss of AOM. Consequently, the maximum theoretical cut-off of the measurement is approximately 2 μm, resulting from the frits that hold the packing material in place at the entrance and exit of the column. The detection limit of the method ranges from 1 to 50 μg C/L depending on the fraction (DOC-Labor, pers. comm.). Feed solutions for the experiments were prepared by diluting the AOM stock solution in synthetic seawater to a biopolymers concentration of 0.5 mg C/L.

5.3.2.2 Biopolymer fractionation

For AOM removal studies, the molecular weight distribution of biopolymers was analyzed with high resolution LC-OCD. In this configuration, separation is done in two columns (Tosoh TSK-HW65S + HW50S) instead of one, allowing for higher resolution of high molecular weight compounds. The HW50S column is only used to separate humic substances and LMW compounds, whereas the separation of high molecular weight material is done on the HW65S column. The method is a very recent development of DOC-Labor and gives semi-quantitative information on four fractions of biopolymers in terms of size, namely, 1000 kDa - 2 μm, 100-1000 kDa, 10-100 kDa, and < 10 kDa. Calibration is done with pullulan standards of known molecular weight. Pullulans are non-ionic extracellular polysaccharides excreted by the fungus Aureobasidium pullulans. Due to their strict linear structure, pullulans are well-defined model substances (Singh et al., 2008) and commonly applied as calibration standards in gel permeation and size exclusion chromatography. After calibration, elution times of 1000 kDa, 100 kDa and 10 kDa are known. Fractions are quantified based on area integration of the chromatogram between known elution times using heart-cuts covering about 90% of standard peak area.

5.3.3 Coagulation process modes

This work investigated the effect of coagulation mode and conditions on UF fouling potential of AOM as measured by MFI-UF, pressure build-up during filtration at constant flux, and AOM removal rates. Experimental conditions for each mode are presented in Table 5-1. Experiments are divided into two parts,

1. Fouling potential measurements and test runs with capillary UF membranes performed under conditions simulating inline coagulation prior to ultrafiltration.

2. Removal of AOM for three coagulation modes, namely,
   - Coagulation, flocculation followed by sedimentation
   - Coagulation, flocculation followed by sedimentation and filtration (0.45 μm)
   - Inline coagulation followed by ultrafiltration (150 kDa).

Table 5-1 Experimental conditions for coagulation modes tested for AOM removal

Mode	Description	Mixing		Flocculation		Settling	Filtration
		G [s^{-1}]	t [s]	G [s^{-1}]	t [min]	[min]	
A	Coag/Flocc/Sed	1100	20	45	15	20	-
B	Coag/Flocc/Sed/Filt	1100	20	45	15	20	0.45 μm
C	Inline coagulation/UF	1100	20	-	-	-	150 kDa

For fouling potential studies, stock AOM suspension was diluted with synthetic seawater (TDS 35 g/L; pH 8.0 ± 0.1; alkalinity 2.5 mmol/L) to a biopolymer concentration of 0.5 mg C/L. This concentration of biopolymers was encountered in North Sea water during algal bloom events that coincided with very high fouling potential in UF systems (Schurer et al., 2013). To investigate the effect of various coagulation conditions such as dose, mixing intensity (G), mixing time and flocculation on the fouling potential of AOM solutions, batch inline coagulation/UF experiments were carried out at different flux rates. Coagulation was done with a calibrated jar tester (ECEngineering Model CLM4, Canada). Ferric chloride was added to 1 L solutions of AOM in synthetic seawater at the start of the rapid mixing step. After coagulant addition, iron hydrolysis is instantaneous and pH decreases immediately; by as much as 1 pH unit for the higher coagulant doses. To maintain a pH range after coagulant addition of 7 - 8 for the experiments, pH correction was performed only for coagulation at 10 mg Fe(III)/L (Table 5-2). pH range 7 - 8 was chosen for the experiments as more and more SWRO plants operate in this range. The jar tester is calibrated for a volume of 1 L, allowing conversion of paddle revolutions (rpm) to mixing intensities (G values) with a maximum intensity of 1100 s^{-1}.

Table 5-2 Initial and final pH values for the range of coagulant dose applied

Dose [mg Fe(III)/L]	$pH_{initial/corrected}$	pH_{final}
0.1	8.0	8.0
0.5	8.0	8.0
1.0	7.9	7.7
2.0	7.9	7.6
5.0	7.9	7.1
10.0	8.6*	7.4

* pH correction was applied only for 10 mg Fe(III)/L to ensure that coagulation is performed at pH range 7-8 for all experiments

To simulate inline coagulation (coagulation mode C), coagulated samples were collected immediately after rapid mixing and filtered through 150 kDa PES membranes (Pentair X-flow,

The Netherlands) operated in inside-out configuration in dead-end mode. In practice 150 kDa PES membranes in inside-out configuration operated in dead-end mode are commonly applied in SWRO pretreatment.

Removal of AOM was studied for three coagulation modes as described in Table 5-1. These modes were designed to simulate different forms of coagulant application in practice. The same coagulation setup was used to perform coagulation in modes A and B for 1 L samples of AOM in synthetic seawater (0.5 mg C/L as biopolymers) at different coagulant dose. Flocculation time was set at 15 min of slow mixing (45 s^{-1}). After flocculation the paddles were gently removed and solutions were allowed to settle for 20 minutes. Samples of the supernatant were collected for analysis from 10 cm below water surface by a tube and syringe assembly to avoid drawing settled or floating flocs. Coagulation mode C was performed as described above for inline coagulation/UF and permeate samples were collected for analysis. Iron residual in permeate samples was measured with the colorimetric phenanthroline method (Sandell, 1959) with limit of detection of 20 μg Fe(III)/L.

5.3.4 Ultrafiltration experiments

The MFI-UF setup (Salinas-Rodriguez, 2011) was adapted to fit a UF pen module - equipped with capillary membrane fibres with an internal diameter of 0.8 mm. Effective filtration area of the UF modules was 0.0021 m^2 ± 5%. Ultrafiltration pen modules were prepared in the laboratory by potting 6 UF hollow fibres (Pentair X-flow, The Netherlands) of 15 cm effective filtration length in transparent flexible polyethylene tubing, PEN-x1, 25-NT (Festo, Germany) using polyurethane glue (Bison, The Netherlands).

Pressure was recorded with a seawater resistant pressure sensor, Cerabar PMC 55 (Endress and Hauser, Switzerland); operational pressure range 0 - 4 bar, maximum deviation 0.04%. Pressure recordings were logged in a computer by connecting the pressure transmitter via a modem, FXA195 Hart (Endress and Hauser, Switzerland) through a USB connection. Data acquisition software, Rensen OPC Office Link (OPC Foundation, USA), was used to log pressure values at a frequency of 10 seconds in MS Excel for further processing.

5.4 Results and discussion

5.4.1 Characterization of harvested AOM solution

Based on LC-OCD results (Table 5-3), AOM harvested from the marine diatom *Chaetoceros affinis* mostly comprised biopolymers, i.e., 51% of the chromatographable organic carbon (COC). The main fraction of the biopolymer peak is polysaccharide-like material. Protein concentration was calculated based on the assumption that all organic nitrogen in the biopolymers fraction originates from proteins. Signals were observed in the UV and organic nitrogen detectors for the biopolymers fraction. These signals may originate partly or completely from stray light effects of colloids/particles in both detectors as sample was not pre-filtered and do not represent aromatic substances.

Table 5-3 LC-OCD analysis of AOM stock solution harvested from Chaetoceros affinis

Fraction	Molecular Weight [Da]	Concentration [mgC/L]	Percentage [%]
COC		10.59	100
Biopolymers	>> 20,000	5.40	51
Polysaccharides		3.62	
Proteins		1.78	
Humic Substances	~ 1,000	2.61	24
Building Blocks	300 - 500	0.36	3
LMW Neutrals	< 350	2.40	22
LMW Acids	< 350	n.q.	n.q.

For the column configuration applied in this measurement, biopolymers elute approximately 25 minutes after sample injection into the column (Figure 5-1). Compounds eluting between approximately 35 and 50 minutes are formally assigned to fractions of humic substances. The compounds do not necessarily fit the strict definition of DOC-Labor for humic substances (based on calibration with Suwannee River humic acids) in terms of molecular weight and aromaticity, but they are certainly humic-like.

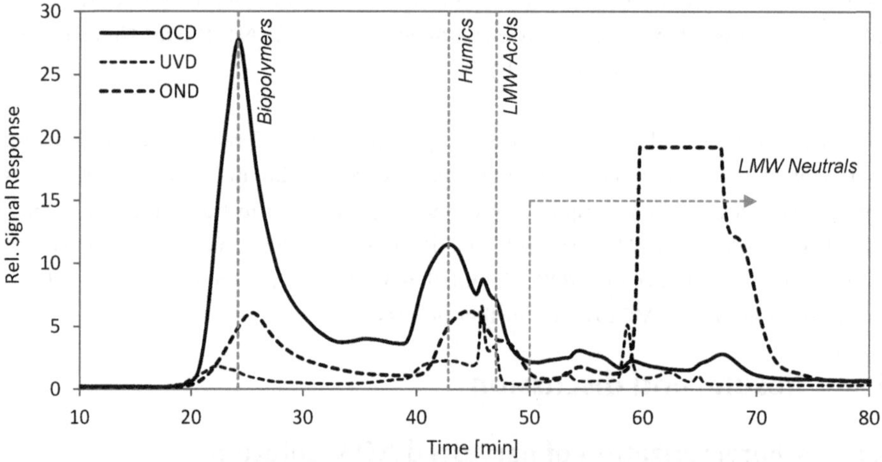

Figure 5-1 OCD-UVD-OND chromatogram of AOM stock solution indicating different fractions of organic carbon and their respective elution time

As AOM stock solutions were not pre-filtered, cell fragments and other debris may be present, resulting in the humic-like substances peak. The concentration of humic-like substances and LMW neutrals may also be attributed to specific compounds in the medium in which the diatoms were cultivated. Diatoms were inoculated in medium containing 0.5 mg/L each of vitamin B12 and biotin. Concentration of these compounds in the AOM stock solution is not known. Some algae release vitamin B12 binding factors (Droop, 1968) during growth, while several species release thiamine and biotin into the medium (Carlucci and Bowes, 1970). Cyanocobalamin was used as Vitamin B12 in this study, with a molecular size of 1,355 Da, indicating that this compound is likely to elute within the humic substances category. Molecular weight of biotin is 244 Da, for which elution is likely to occur in the category of LMW neutrals and LMW acids.

5.4.2 Effect of inline coagulation conditions on AOM fouling potential

This section investigates the effect of coagulation conditions and filtration flux on the fouling potential of AOM obtained from *Chaetoceros affinis* in UF membranes.

5.4.2.1 Dose

The effect of coagulant dose on fouling potential of AOM in synthetic seawater was evaluated at constant rapid mixing conditions of 1100 s^{-1} for 20 seconds, constant filtration flux and pH values given in Table 5-2. Two distinct aspects of the effect of coagulant dose were identified, (i) fouling potential (as measured by MFI-UF constant flux) and (ii) compressibility (as observed from TMP profiles in ultrafiltration at constant flux).

Fouling potential of coagulated AOM was quantified in terms of MFI-UF for filtration through 150 kDa PES membrane fibres at 100 L/m^{2}h (Figure 5-2). Coagulant dose of 0.1 mg Fe(III)/L had no affect on AOM fouling potential, while the effect of coagulation at 0.5 mg Fe(III)/L was marginal. MFI-UF was substantially lower at coagulant dose ≥ 1 mg Fe(III)/L. Considering the definition of MFI-UF and specific cake resistance (Eq. 5-5, Eq. 5-6), reduction in MFI-UF may be due to an increase in aggregate size, resulting in higher cake/gel porosity. Aggregates formed by coagulation are fractals - self repeating structures whose properties change with change in size; aggregate porosity increases with size (Jiang and Logan, 1991). Aggregates with high porosity form porous deposits with low specific cake resistance and consequently low MFI-UF values.

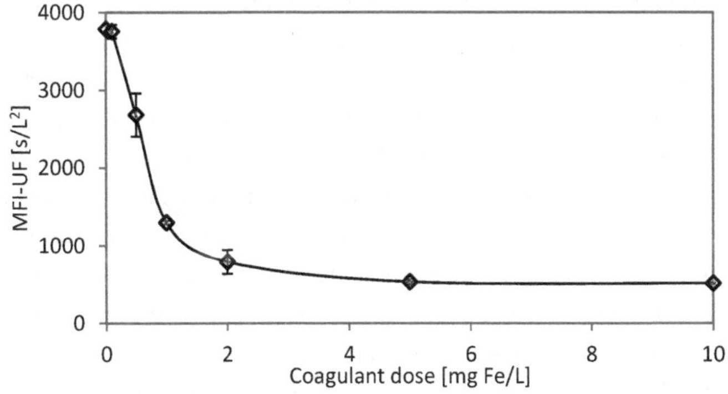

Figure 5-2 Effect of coagulant dose on MFI-UF of coagulated AOM obtained from *Chaetoceros affinis* (0.5 mg C/L as biopolymers) for mixing at 1100 s^{-1} for 20 sec

AOM filtration in the absence of coagulant is marked by a notable curvature, which is reduced and gradually diminished, rendering linear curves, with increase in coagulant dose (Figure 5-3a). The slopes of the TMP versus time curves further validate the change in curvature with increasing coagulant dose (Figure 5-3b). For slopes calculated from data sets of 5 minutes moving average, the minimum slope (from which MFI-UF is calculated) occurs in the first 10 minutes of filtration, regardless of coagulant dose. In the absence of coagulant or at low doses of up to 0.5 mg Fe(III)/L, slope increases with filtration time. However, when coagulant dose is increased to 1 mg Fe(III)/L, the slope is constant throughout the filtration cycle.

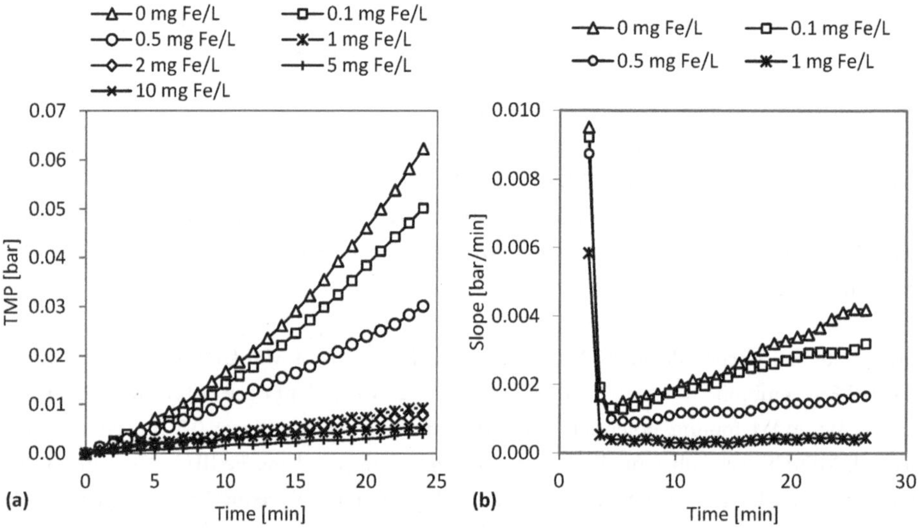

Figure 5-3 (a) Filtration curves and (b) slope development for coagulated AOM obtained from *Chaetoceros affinis* (0.5 mg C/L as biopolymers) at 1100 s^{-1} for 20 s, at constant filtration flux of 100 L/m^2h through 150kDa PES membranes

The marked curvature in TMP profiles for AOM filtration may be due to enhanced compression of the cake/gel layer of AOM on the membrane surface. AOM is amorphous (Passow, 2002) and can be rearranged and deformed into a more compact state as head loss over the cake/gel layer increases due to increase in layer thickness. Salinas Rodriguez (2011) noted that compression/compaction starts almost immediately for heterogeneous cakes of organic and inorganic particles in fresh water and seawater. Alternatively, as smaller AOM gradually reaches the membrane, surface and/or pore blocking occurs causing a reduction in available filtration area. This reduction in area creates an increase in local filtration flux on the remaining available membrane area for filtration, affecting filtration in two ways; (i) as filtration area is reduced due to blocking and filtration flow is kept constant, the local flux rate on the remaining area increases, corresponding with a second order increase in TMP (Eq. 5-4). (ii) as flux affects compression rate (Salinas-Rodriguez, 2011) a higher local flux will result in higher compression of the filter cake and an increasing slope. A third explanation might be that particles that could initially pass through the membrane or cake/gel layer pores, may gradually deposit within the pores or on the surface of the cake/gel layer resulting in pore constriction and reduced permeability. In other words, a transition occurs to filtration through a secondary (virtual) membrane with gradually smaller pores imparting increasingly higher resistance to filtration. Enhanced compression (rearrangement and compaction) of the cake/gel layer is likely to be the main fouling mechanism for AOM in UF membranes.

When coagulant is added to AOM solutions, iron reacts with the biopolymers, changing the properties in such a way that the specific resistance (α) and compressibility of the AOM cake/gel layer are reduced. At 0.5 mg Fe(III)/L and pH 8, the slight reduction in fouling potential and compressibility may be due to the formation of colloidal Fe-biopolymer complexes, as a relatively small amount of positively charged iron species is available for complexation to occur. At 0.1 mg Fe(III)/L, the concentration of positively charged species is too low for any reaction to occur and consequently AOM fouling potential and compressibility are not affected. At coagulant dose ≥ 1 mg Fe(III)/L and pH values of 7-7.7, AOM adsorption on iron hydroxide precipitates takes place, resulting in the formation of Fe-biopolymer aggregates that are relatively large and less compressible. It is clear that at higher coagulant dose, properties of the cake/gel layer trend toward the properties of iron hydroxide precipitates. Fe residual in all permeate samples was below detection limit, eliminating the assumption that soluble Fe-biopolymer complexes were formed.

5.4.2.2 Filtration Flux

To elaborate the effect of flux on ultrafiltration of seawater containing AOM, experiments were performed at different flux values (50-350 L/m^2h) in the absence and presence of coagulant for rapid mixing at 1100 s^{-1} for 20 secs. Filtration volume was kept constant for all experiments i.e., at higher flux values filtration times were shortened.

For AOM filtration in the absence of coagulation (Figure 5-4), MFI-UF was substantially higher at higher filtration flux. This observation is in concert with measurements with real seawater (Salinas Rodriguez et al., 2012). The strong flux dependency of MFI-UF suggests that AOM cake/gel layer is highly compressible. At 5 mg Fe(III)/L, MFI-UF values are low and hardly affected by increase in flux, indicating that aggregates formed from coagulation at 5 mg Fe(III)/L create highly porous cakes. Moreover, MFI-UF values at this dose are very close to measured MFI-UF levels of iron hydroxide precipitates in the absence of turbidity and organic matter.

Filtration curves at different flux values without coagulation, and with 1 and 5 mg Fe(III)/L are presented in Figure 5-5. As expected, at higher flux overall pressure increase and the rate of pressure development in one cycle are higher.

Based on the observed effects of coagulant dose and filtration flux on fouling potential and pressure development, it may be concluded that for a given rate of pressure development, higher coagulant dose allows for filtration at higher flux. This is illustrated in Table 5-4, e.g., pressure builds up at the rate of 0.022 bar/hr for direct filtration of AOM at 50 L/m²h when no coagulant is added. To maintain a similar rate of pressure development at much higher filtration flux, e.g., 100 L/m²h, coagulant dose of 1 mg Fe(III)/L is required.

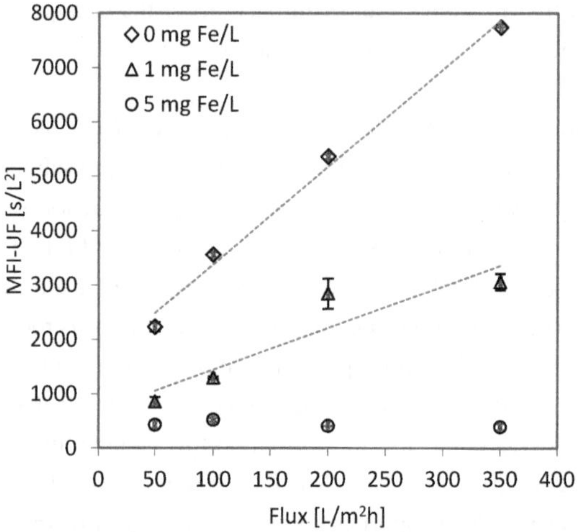

Figure 5-4 Effect of flux on fouling potential of AOM (0.5 mg C/L as biopolymers) at fixed mixing conditions and filtration through 150 kDa PES membranes

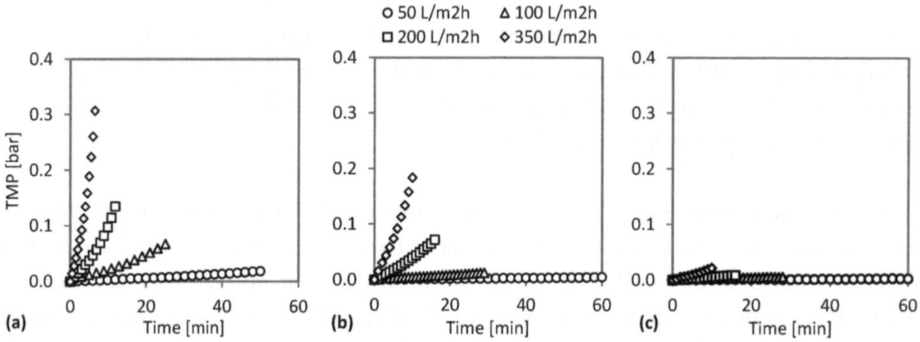

Figure 5-5 TMP increase for AOM filtration at different flux values for (a) no coagulant, (b) 1 mg Fe(III)/L and (c) 5 mg Fe(III)/L

Table 5-4 Average filtration slope as a function of coagulant dose at different flux values

Flux	Filtration Slope [bar/hr]		
[L/m²h]	0 mg Fe(III)/L	1 mg Fe(III)/L	5 mg Fe(III)/L
50	0.022	0.004	0.002
100	0.150	0.024	0.010
200	0.680	0.248	0.033
350	3.145	1.142	0.130

5.4.2.3 Mixing

To verify the effect of mixing intensity and time on inline coagulation, G-values of 100 and 1100 s^{-1} were applied for 20 and 240 secs. MFI-UF was measured by filtering the solutions through 150 kDa membranes at a constant flux of 100 L/m^{2}h.

At a coagulant dose of 1 mg Fe(III)/L (Figure 5-6), mixing intensity of 1100 s^{-1} resulted in substantially lower MFI-UF values. No difference was observed between mixing time of 20 and 240 seconds. At 1 mg Fe(III)/L and pH 7.7, a relatively small amount of positively charged iron species are present (Edzwald and Haarhoff, 2011) for precipitation of colloidal Fe-biopolymer complexes. These species are formed in < 0.1 s (Edzwald et al., 1998) if no hydroslysis polymers are formed and within 1 s if polymers are formed (Hahn and Stumm, 1968; Johnson and Amirtharajah, 1983). Higher G-values result in shorter transport time of reactive iron species to particles and colloids in the water before reactivity is reduced due to polymerization and hydroxide precipitation. These considerations suggest that under such conditions, positively charged iron species react with the negatively charged biopolymers to a certain extent.

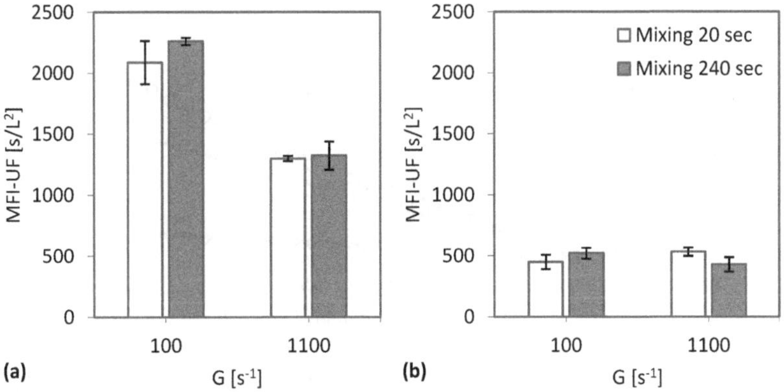

Figure 5-6 Effect of mixing intensity (G) and mixing time on fouling potential of coagulated AOM for (a) 1 mg Fe(III)/L and (b) 5 mg Fe(III)/L

At 5 mg Fe(III)/L, MFI-UF values were not influenced by mixing intensity and mixing time and fouling potential was dominated by the characteristics of iron hydroxide precipitates. At this dose and pH, the mechanisms governing coagulation is adsorption to and enmeshment in iron hydroxide precipitates during sweep floc, resulting in the formation of relatively large Fe-biopolymer aggregates. As such, time for coagulant delivery is not crucial. MFI-UF values for iron hydroxide precipitates in the absence of turbidity and organic matter were measured in the range of 200 - 500 s/L^2 for various coagulation conditions (data not shown).

5.4.3 Prediction of TMP increase in UF membranes

Reduction in compressibility of the AOM cake/gel layer is noticeable even at the lowest coagulant concentration of 0.1 mg Fe(III)/L (Figure 5-3). However, MFI-UF values for the same experimental conditions did not reflect this difference (Figure 5-2). This implies that prediction - in this case - of pressure increase in one cycle cannot be done with MFI-UF values alone as compression effects that dictate the total pressure increase in one cycle are not reflected by these measurements.

Measured and predicted TMP profiles in a filtration cycle for no coagulation, 0.5 mg Fe(III)/L and 5 mg Fe(III)/L are presented in Figure 5-7. Predicted TMP values were lower - as expected - than the measured TMP for both cases.

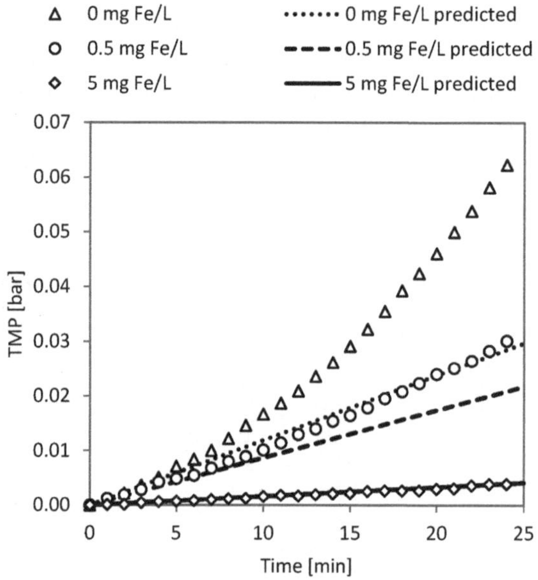

Figure 5-7 Predicted and measured TMP profiles for AOM filtration in the absence and presence of coagulant

The difference between measured and predicted TMP in the absence of coagulation was 200% at the end of the filtration cycle of 25 minutes. This difference dropped to approximately 50% for coagulation at 0.5 mg Fe(III)/L.

At coagulant concentrations of 1 mg Fe(III)/L and above, measured and predicted pressure increase throughout a filtration cycle were almost identical. This indicates that simple cake/gel filtration mechanism is predominant for coagulation at these concentrations.

5.4.4 AOM removal by different coagulation modes

AOM removal was quantified through normalized removal of biopolymers for three coagulation modes A, B, C as described in section 3.3.

5.4.4.1 Organic carbon fractionation and biopolymer removal

Chromatograms obtained from LC-OCD analysis of the feed solution before and after pretreatment with coagulation/flocculation/sedimentation (mode A) and inline coagulation/UF (mode C) are presented in Figure 5-8. Samples were analyzed with a special column combination that allows for higher resolution of larger molecular weight compounds. The later elution time of biopolymers (~ 60 min) in this column configuration is due to longer retention time resulting from the application of two columns. Reduction in organic carbon concentration was largely due to a reduction in biopolymers concentration (peak X) as the low molecular weight acids fraction (peak Y) was not affected by coagulation mode and/or coagulant dose.

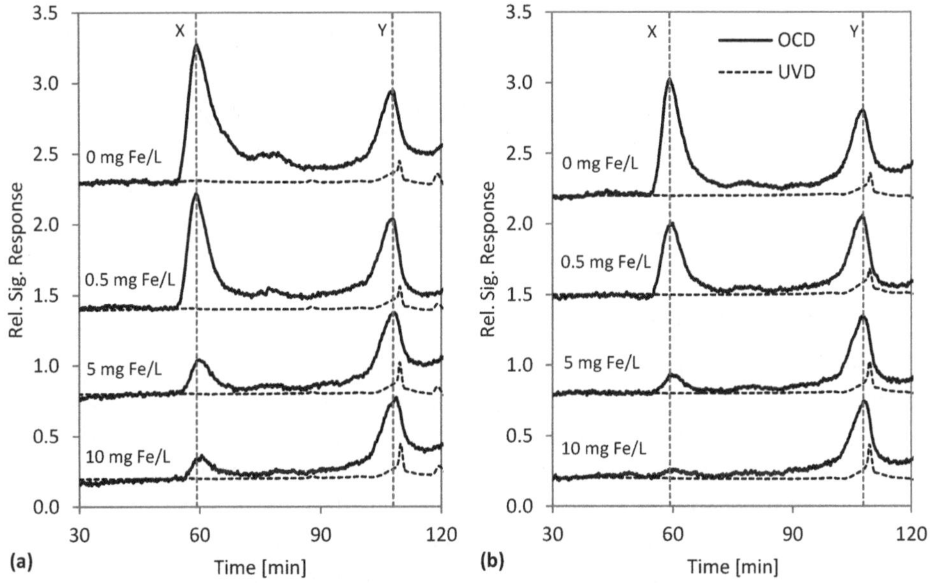

Figure 5-8 OCD-UVD chromatograms of AOM solutions coagulated with $FeCl_3$ for (a) coagulation/flocculation followed by settling and (b) inline coagulation/UF; peak X corresponds to biopolymers elution at approximately 60 minutes, peak Y corresponds to LMW acids elution at approximately 110 minutes. Offsets in signal response were made for clarity

The removal of biopolymers was investigated for different coagulation modes (Figure 5-9) as a function of coagulant dose. When no coagulant was added, mode C had the highest removal of biopolymers followed by mode B and mode A. For mode A, the reduction in biopolymers in the absence of coagulant could be due to the aggregation of small colloidal biopolymers that enhances their settling. Bernhardt et al. (1985) suggested that extracellular dissolved organic matter aid flocculation by bridging at appropriate surface coverage.

Coagulation at 0.5 mg Fe(III)/L enhanced biopolymer removal for all coagulation modes. At 5 and 10 mg Fe(III)/L, the difference in removal rates for different coagulation modes became negligible. Increase in dose shifts the predominant coagulation mechanism to sweep floc and inter-particle bridging, whereby removal is enhanced by adsorption to and/or enmeshment in precipitated iron hydroxide.

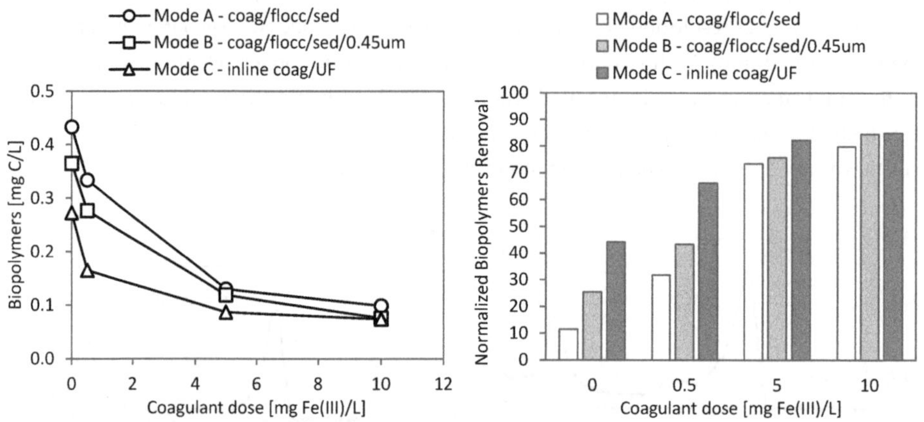

Figure 5-9 Total biopolymer concentration and normalized biopolymer removal as a function of coagulant dose for different coagulation modes

5.4.4.2 Biopolymer Fractionation

High resolution LC-OCD was performed for better separation of biopolymers from humic substances. This method is a very recent development of DOC-Labor and provides semi-quantitative information on four fractions of biopolymers in terms of molecular weight, namely, 1000 kDa - 2 μm, 100-1000 kDa, 10-100 kDa, < 10 kDa. Fractions are quantified based on area integration (area boundaries are defined by pullulan standards) using heart-cuts covering about 90% of standard peak area. However, resolution is poor for fraction < 10 kDa, as this fraction is superimposed by the building blocks and low molecular weight acids fractions (Figure 5-10). For this fraction the quantification is rather arbitrary and bias may well exceed 50 %.

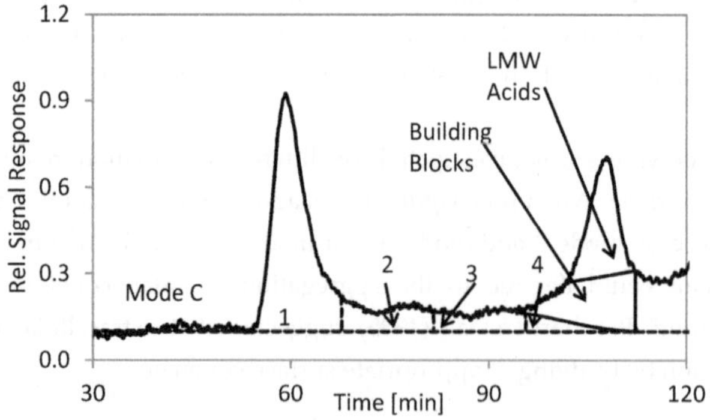

Figure 5-10 Division of OCD chromatograms based on calibration with pullulan standards to obtain various fractions of biopolymers based on molecular weight

Concentrations of different biopolymer fractions as a function of coagulation mode and coagulant dose are summarized in Figure 5-11. Biopolymers were mainly composed of larger molecular weight fractions; approximately 80% of the total biopolymer concentration was larger than 100 kDa. As a consequence, reduction in biopolymer concentration was mainly due to the removal of biopolymers > 100 kDa. Coagulation had a strong impact on the removal of biopolymers larger than 100 kDa for all modes of coagulant application where removal was substantially higher at higher coagulant dose.

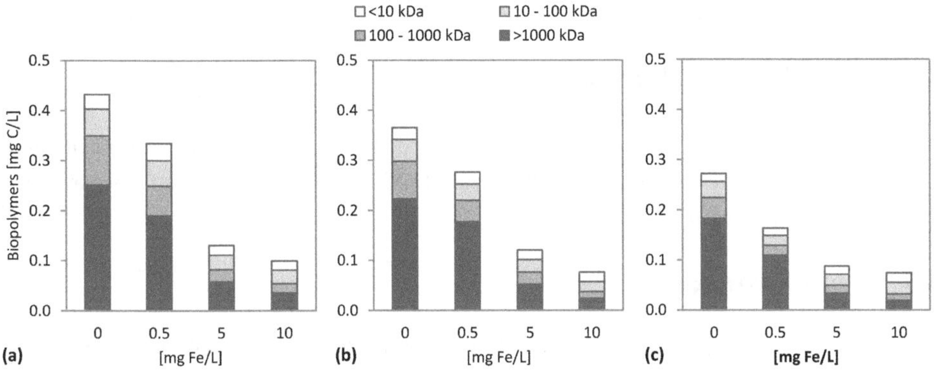

Figure 5-11 Biopolymer concentrations for fractions of different molecular weight as a function of coagulant dose for (a) mode A, (b) mode B, and (c) mode C

5.5 Conclusions

Algal organic matter harvested from the marine diatom *Chaetoceros affinis* was mainly composed of biopolymers. AOM had high fouling potential as measured by MFI-UF and was very compressible. Filtration at higher flux exacerbated both fouling potential and compressibility of AOM. AOM is amorphous and deformable and can form cake/gel layers that are rearranged and compacted at higher filtration flux resulting in higher pressure development.

Coagulation substantially reduced fouling potential and compressibility of AOM. At coagulant dose of 1 mg Fe(III)/L and higher, AOM fouling characteristics diminished and those of iron hydroxide precipitates prevailed, indicating that the predominant coagulation mechanism is adsorption of biopolymers on and enmeshment in iron hydroxide precipitates forming Fe-biopolymer aggregates.

At low coagulant dose, fouling potential and compressibility were marginally improved as coagulation most likely resulted in partial complexation of the biopolymers and formation of colloidal Fe-biopolymer complexes. Coagulation also substantially reduced the flux dependency of AOM filtration, resulting in substantially lower MFI-UF values and linear development of pressure in filtration tests at constant flux. For coagulation at 1 mg Fe(III)/L, coagulant delivery time was crucial, further suggesting that at low coagulant concentration reactions between positively charged iron species and negatively charged biopolymers occurs to a certain extent.

AOM removal was mainly due to the removal of biopolymers as low molecular weight acids were not removed for different coagulation modes and coagulant dose. High resolution LC-OCD showed that biopolymers were mainly composed of higher molecular weight fractions (> 100 kDa). Inline coagulation/UF (mode C) was most effective in terms of biopolymer removal at low coagulant dose. However, at high coagulant dose, biopolymer removal was dominated by coagulant dose rather than coagulation mode as sweep floc becomes the dominant mechanism.

5.6 Acknowledgements

Thanks are due to CCAP, Scotland for providing the diatom strain, L. Villacorte for cultivating the marine diatoms, Dr. Huber and Dr. Balz from DOC-Labor for LC-OCD analyses and data interpretation, M. Joulie and M. Gonzalez.

5.7 References

Berman, T., Holenberg, M., 2005. Don't fall foul of biofilm through high TEP levels. Filtration + Separation 42, 30–32.

Berman, T., Mizrahi, R., Dosoretz, C. G., 2011. Transparent exopolymer particles (TEP): A critical factor in aquatic biofilm initiation and fouling on filtration membranes. Desalination 276(1-3), 184-190.

Bernhardt, H., Hoyer, O., Schell, H., Lusse, B., 1985. Reaction mechanisms involved in the influence of algogenic matter on flocculation. Zeitschrift fur Wasser und Abwasser forschung 18, 18-30.

Boerlage, S.F.E., Kennedy, M., Tarawneh, Z. Faber, R.D., Schippers, J.C., 2004. Development of the MFI-UF in constant flux filtration. Desalination 161, 103–113.

Busch, M., Chu, R. & Rosenberg, S., 2010. Novel trends in dual membrane systems for seawater desalination: minimum primary pre-treatment and low environmental impact treatment schemes. IDA Journal, Desalination and Water Reuse, 2(1), pp.56-71.

Carlucci, A.F., Bowes, P.M., 1970. Vitamin production and utilization by phytoplankton in mixed culture. J. Phycology 6, 393-400.

Cho, M.H., Lee, C.H., Lee, S., 2006. Effect of flocculation conditions on membrane permeability in coagulation - microfiltration. Desalination 191, 386-396.

Choi, K.Y., Dempsey, B.A., 2004. Inline coagulation with low pressure membrane filtration. Water Research 38, 4271-4281.

Dam, H.G., Drapeau, D.T., 1995. Coagulation efficiency, organic-matter glues and the dynamics of particles during a phytoplankton bloom in a mesocosm study. Deep-Sea Research II 42(1), 111-123.

Dempsey, B.A., Ganho, R.M., O'Melia, C.R., 1984. The coagulation of humic substances by means of aluminium salts. Journal of the American Water Works Association 76(4), 141–150.

Dennett, K.E., Amirtharajah, A., Moran, T.F., Gould, J.P., 1996. Coagulation: its effect on organic matter. Journal of the American Water Works Association 84(4), 129–142.

Droop, M.R., 1968. Vitamin B12 and marine ecology. IV. The kinetics of uptake, growth and inhibition in *Monochrysis lutheri*. J. Mar. Biol. Assoc. UK 48, 689-733.

Duan, J., Graham, N.J.D., Wilson, F., 2002. Coagulation of humic acid by ferric chloride in saline (marine) water conditions. Water Science and Technology 47(1), 41-48.

Edzwald, J.K., Haarhoff, J., 2011. Seawater pretreatment for reverse osmosis: Chemistry, contaminants, and coagulation. Water Research 45, 5428-5440.

Edzwald, J.K., Janssens, J.G., Wiesner, M.R., 1998. Process selection considerations. In: McEwen, J.B. (Ed.) Treatment process selection for particle removal. Colorado: AwwaRF and AWWA.

Fogg, G.E., 1983. The ecological significance of extracellular products of phytoplankton photosynthesis. Bot. Marina 26, 3–14.

Gray, K.A., 1988. The preparation, characterisation and use of inorganic iron (III) polymers for coagulation in water treatment. Doctoral dissertation, The John Hopkins University, Baltimore, MD.

Guigui, C., Rouch, J.C., Durand-Bourlier, L., Bonnelye V., Aptel, P., 2002. Impact of coagulation conditions on the inline coagulation/UF process for drinking water production. Desalination 147, 95-100.

Hahn, H.H., Stumm, W., 1968. Kinetics of coagulation with hydrolysed Al (III). Journal of Colloid Interface Science 28, 134-144.

Hellebust, J.A., 1965. Excretion of some organic compounds by marine phytoplankton. Limnology and Oceanography 10, 192-206.

Henderson, R.K., Baker, A., Parsons, S.A., Jefferson, B., 2008. Characterisation of algogenic organic matter extracted from cyanobacteria, green algae and diatoms. Water Research 42, 3435-3445.

Howe, K.J., Clark, M.M., Effect of coagulation pretreatment on membrane filtration performance. J. AWWA 98(4), 133–146.

Huber, S.A., Balz, A., Abert, M., Pronk, W., 2011. Characterisation of aquatic humic and non-humic matter with size-exclusion chromatography - organic carbon detection - organic nitrogen detection (LC-OCD-OND). Water Research 45, 879-885.

Huber, S.A., Frimmel, F.H., 1994. Characterization and quantification of marine dissolved organic carbon with a direct chromatographic method. Environmental Science and Technology 28, 1194-1197.

Jarvis, P., Jefferson, B., Parsons, S.A., 2006. Floc structural characteristics using conventional coagulation for a high doc, low alkalinity surface water source. Water Research 40, 2727-2737.

Jermann, D., Pronk, W., Kagi, R., Halbeisen, M., Boller, M., 2008. Influence of interactions between NOM and particles on UF fouling mechanisms. Water Research 42, 3870-3878.

Jiang Q., Logan B.E., 1991. Fractal dimensions of aggregates determined from steady-state size distribution. Environmental Science and Technology 25, 2031–2038.

Johnson, P.N., Amirtharajah, A., 1983. Ferric chloride and alum as single and dual coagulants. Journal of American Water Works Association 75 (5), 232-239.

Kimura, K., Tanaka, K., Watanabe, Y., 2014. Microfiltration of different surface waters with/without coagulation: Clear correlations between membrane fouling and hydrophilic biopolymers. Water Research 49, 434-443.

Lee, J.D., Lee, S.H., Jo, M.H., Park, P.K., Lee, C.H., Kwak, J.W., 2000. Effect of coagulation conditions on membrane filtration characteristics in coagulation-microfiltration process for water treatment, Environmental Science and Technology 17, 3780-3788.

Lyman, J., Fleming, R.H., 1940. Composition of seawater. Journal of Marine Research 3, 134-146.

Mopper, K., Zhou, J., Sri Ramana, K., Passow, U., Dam, H.G., Drapeau, D.T., 1995. The role of surface-active carbohydrates in the flocculation of a diatom bloom in a mesocosm. Deep-Sea Research Part II 42(1), 47-73.

Myklestad S., 1974. Production of carbohydrates by marine planktonic diatoms, 1 - comparison of nine different species in culture. Journal of Experimental Marine Biology and Ecology. 15, 261-274.

Myklestad, S., 1995. Release of extracellular products by phytoplankton with special emphasis on polysaccharides. Science of the Total Environment 165, 155-164.

Passow, U., 2002. Transparent exopolymer particles (TEP) in aquatic environments. Progress in Oceanography 55(3), 287-333.

Passow, U., Alldredge, A. L., 1994. Distribution, size, and bacterial colonization of transparent exopolymer particles (TEP) in the ocean. Marine Ecology Progress Series 113, 185–198.

Pearce, G., 2009. SWRO pre-treatment: treated water quality. Filtration+Separation 46(6), 30-33.

Pearce, G.K., 2007. The case for UF/MF pre-treatment to RO in seawater applications. Desalination 203(1-3), 286-295.

Qu, F., Liang, H., He, J., Ma, J., Wang, Z., Yu, H., Li, G., 2012a. Characterization of dissolved extracellular organic matter (dEOM) and bound extracellular organic matter (bEOM) of microcystis aeruginosa and their impacts on UF membrane fouling. Water Research 46(9), 2881-2890.

Qu, F., Liang, H., Tian, J., Yu, H., Chen, Z., Li, G., 2012b. Ultrafiltration (UF) membrane fouling caused by cyanobateria: Fouling effects of cells and extracellular organics matter (EOM). Desalination 293, 30-37.

Rebhun, M., Lurie, M., 1993. Control of organic matter by coagulation and floc separation. Water Science and Technology 27 (11), 1-20.

Ruth, B.F., Montillon, G.H., Montana, R.E., 1933. Studies in filtration: Part I, critical analysis of filtration theory. Ind Eng Chem 25, 76-82.

Salinas Rodriguez, S.G., 2011. Particulate and organic matter fouling of SWRO systems: characterization, modelling and applications. Doctoral dissertation, UNESCO-IHE/TU Delft, Delft.

Salinas Rodriguez, S.G., Kennedy, M.D., Amy, G.L., Schippers, J.C., 2012. Flux dependency of particulate/colloidal fouling in seawater reverse osmosis systems. Desalination and Water Treatment 42, 155-162.

Sandell, E.B., Colorimetric Determination of Traces of Metals, 3rd Ed. Interscience Publishers, Inc., New York, 1959.

Schippers, J.C., Verdouw, J., 1980. The modified fouling index, a method of determining the fouling characteristics of water. Desalination 32, 137-148.

Schurer R., Tabatabai A., Villacorte L., Schippers J.C., Kennedy M.D., 2013. Three years operational experience with ultrafiltration as SWRO pre-treatment during algal bloom. Desalination and Water Treatment 51(4-6), 1034-1042.

Singh, R.S., Saini, G.K., Kennedy, J.F., 2008. Pullulan: Microbial sources, productions and applications. Carbohydrate Polymers 73, 515-531.

Tian, J., Ernst, M., Cui, F., Jekel, M., 2013. Correlations of relevant membrane foulants with UF membrane fouling in different waters. Water Research 47, 1218-1228.

Villacorte, L.O., 2014. Algal blooms and membrane based desalination technology. Doctoral dissertation, UNESCO-IHE/TU Delft, Delft.

Villacorte, L.O., Ekowati, Y., Winters, H., Amy, G.L., Schippers, J.C., Kennedy, M.D., 2013. Characterisation of transparent exopolymer particles (TEP) produced during algal bloom: a membrane treatment perspective. Desalination and Water Treatment 51(4-6), 1021-1033.

Villacorte, L.O., Schurer, R., Kennedy, M., Amy, G., Schippers, J.C., 2010. The fate of transparent exopolymer particles in integrated membrane systems: a pilot plant study in Zeeland, The Netherlands. Desalination and Water Treatment 13, 109-119.

Watt, W.D., 1968. Extracellular release of organic matter from two freshwater diatoms. Annals of Botany 33, 427-37.

Wilf, M., Schierach, M.K., 2001. Improved performance and cost reduction of RO seawater systems using UF pre-treatment. Desalination 135, 61-68.

Ye, Y., Le Clech, P., Chen, V., Fane, A., Jefferson, B., 2005. Fouling mechanisms of alginate solutions as model extracellular polymeric substances. Desalination 175, 7–20.

6

COATING ULTRAFILTRATION MEMBRANES IN SEAWATER REVERSE OSMOSIS PRETREATMENT

Previous research has shown that pre-coating ultrafiltration (UF) membranes with a removable layer of particles at the start of each filtration run is more capable than inline coagulation in mitigating fouling for operation on low salinity surface water. This chapter investigated the applicability of this method to seawater with high fouling propensity, i.e., algal bloom-impacted seawater. To prove the principle in seawater, UF membranes were coated with iron (hydr)oxide particles at the start of each filtration cycle at different concentrations. Algal organic matter, with biopolymer concentration of 0.2 to 0.7 mg C/L, caused high fouling potential and poor backwashability of the UF membranes. Pre-coating with ferric hydroxide prepared by simple precipitation followed by low intensity grinding was effective in controlling non-backwashable fouling. However, relatively high equivalent dose (~ 3-6 mg Fe(III)/L) was applied. Reducing particle size to the range of 1 μm - prepared by precipitation followed by high intensity grinding - resulted in low equivalent dose (~ 0.3 to 0.5 mg Fe(III)/L) required for stable operation of the UF membranes. Coating improved UF operation through a combination of physical and chemical interactions between coating material and the biopolymers attached to the membrane surface. Coating layers provide a physical barrier to avoid attachment of sticky biopolymers to the membrane surface. At the same time ferric hydroxide particles are assumed to interact with the foulant layer in such a way that the backwashing process is further enhanced.

6.1 Introduction

Organic matter released into seawater by phytoplanktons during algal bloom periods can seriously deteriorate the performance of UF membranes as pretreatment to SWRO systems (Caron et al., 2010). Algal organic matter (AOM) is released copiously by algae throughout their entire life cycle and is comprised of extracellular and intracellular organic matter. The extracellular fraction of AOM is produced by metabolic excretion (Fogg, 1983) and is composed mainly of polysaccharides (Myklestad et al., 1972). Diatoms are particularly known for excreting large quantities of polysaccharides (Watt, 1969) including transparent exopolymer particles (TEP) (Myklestad, 1995; Villacorte, 2013).

TEP are amorphous, gel-like, fractal particles that vary in size up to a few hundred microns (Passow and Alldredge, 1994). They are highly sticky, owing to the sulfate half-ester groups that functionalize their surface (Zhou et al., 1998), resulting in TEP being 2-4 orders of magnitude stickier than other particles in seawater (Passow et al., 1994). The high stickiness of TEP can cause these particles to strongly attach to the surface of MF/UF membranes or within the pores, resulting in poor backwashability (Villacorte et al., 2010; Schurer et al., 2013). TEP can be measured by staining with Alcian blue - a dye specific for acidic polysaccharides and glycoproteins (Passow and Alldredge, 1994; Villacorte, 2014). Biopolymer concentration (polysaccharides and proteins) measured by liquid chromatography - organic carbon detection (LC-OCD) can be used as a surrogate for quantifying AOM concentration in seawater.

To avoid interaction between sticky algal biopolymers and the membrane, either the AOM or the membrane has to be modified. AOM modification can be achieved by common physico-chemical processes such as coagulation. Coagulation with e.g., ferric hydroxide can enhance UF operation by reducing the extent of non-backwashable fouling through complexation of AOM by precipitation and/or adsorption and enmeshment of AOM in precipitated metal hydroxide. Optimizing coagulation prior to UF is not a well-addressed topic and overdosing, underdosing and inadequate mixing conditions are expected to occur in such systems. Consequently, UF membranes may suffer from fibre plugging due to high coagulant dose, or pore blocking by smaller monomers, dimers, and trimers of the hydrolysing metal coagulant. Fouling by Fe^{2+} and Mn^{2+} may also occur with coagulants of low grade. Fouling by iron requires a specifically tailored and intensive cleaning protocol to recover permeability.

Alternatively, membrane surface can be modified to reduce membrane-foulant interaction. Three general surface modification techniques have been applied to enhance the fouling resistance of water treatment membranes (Figure 6-1). Modified membrane synthesis with fluorinated polymer additives showed enhanced water flux for an oil/water feed, compared to an unmodified sample (Khayet et al., 2002). Free-radical, photochemical-, radiation-, redox- and plasma-induced grafting have all been reported as methods to covalently attach monomers to a membrane surface (Ulbricht & Belfort, 1996) using a variety of hydrophilic methacrylate- and polyethylene glycol-based monomers to various membrane types, and slower flux decline compared to unmodified membranes was reported (Susanto and Ulbricht, 2006). A third surface modification method is physical coating, typically with polymers (Kim et al., 1988; La et al., 2011). Surface modification techniques normally focus on hydrophilization of the membrane surface to reduce hydrophobic interactions (Brink & Romijn, 1990). However, this is not effective when hydrophilic components of organic matter are the major foulants, such as the case of an algal bloom; AOM is characterized as predominantly hydrophilic with low SUVA (Henderson, 2008).

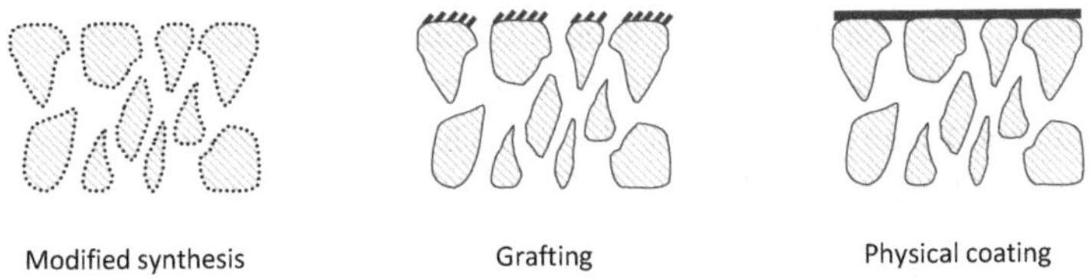

Figure 6-1 Membrane modification techniques to reduce fouling of UF membranes

An alternative anti-fouling approach to reduce membrane-foulant interactions is the application of a removable coating layer to protect the underlying membrane. This method was originally proposed by Galjaard et al. (2001a) and tested in freshwater applications at laboratory and pilot scale. In this approach, the membranes are coated at the start of each filtration cycle, during a very short period to immediately protect the membrane surface. Thereafter, feed water is filtered directly through the coated membrane. The coating layer is expected to shield the membrane surface and reduce contact area between membrane and foulant, thereby reducing permeability loss. At the end of each cycle backwashing is applied to remove both the coating layer and foulants from the membrane surface. The process is schematically shown in Figure 6-2.

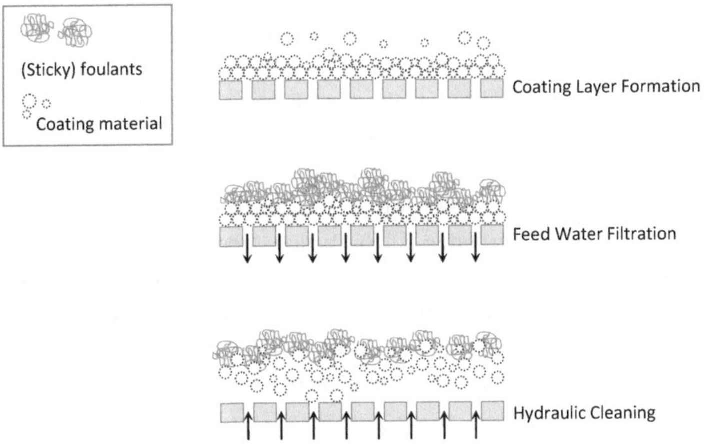

Figure 6-2 Schematic presentation of coating UF membranes to enhance operation (adapted from Galjaard et al., 2001a)

Galjaard et al. (2001a; 2001b; 2003) showed that pre-coating was more successful than inline coagulation with ferric chloride and polyaluminium chloride to control pressure development in UF membranes, in pilot studies. The authors observed that while inline coagulation was incapable of controlling pressure development in the UF system, pre-coating with e.g., ferric hydroxide or diatomite particles, resulted in stable operation. The exact quality composition of the feed waters tested in the studies was not reported.

The success of coating is expected to depend on the creation of a thin layer with relatively low porosity and high permeability that covers the entire membrane surface without blocking the pores. In addition, the coating material should be fully rejected by the MF/UF membranes. Meeting these requirements, this approach will avoid the reported fouling phenomena occurring in inline coagulation (Schurer et al., 2013). The main challenge is to create a stable suspension in a simple and low-cost manner.

6.2 Aim and approach

The aim of this study was to investigate the effect of coating with ferric hydroxide on the performance of capillary UF membranes operating in inside-out mode during algal blooms in seawater. The specific objectives were,

- To validate the proof of principle of coating for UF membranes operating on seawater with substantial AOM concentrations;
- To minimize the amount of coating material required to stabilize UF operation in such conditions.

The following approach was taken to fulfil the objectives,

- To conduct preliminary experiments reproducing the experimental conditions of Galjaard et al. (2001a) on seawater with high AOM concentrations;
- To demonstrate theoretically and experimentally that particle size plays a significant role in the success of coating;
- To elaborate on the mechanisms involved in UF performance enhancement with coating.

Suspensions of iron hydroxide particles created by chemical precipitation were subjected to grinding at various intensities to create different particle size distributions. Particles were characterized by dynamic light scattering (DLS). Particle deposition and uniformity of coverage in a single UF capillary was investigated for particle suspensions prepared at different grinding intensity by measuring ferric concentration along the capillary and by visual observation of autopsied capillaries. Finally, performance of UF membranes coated with suspensions of different particle size was investigated at for filtration of feed water with high AOM content.

The main assumption in this study is that full coverage of the membrane surface by the coating material is essential to avoid attachment of sticky algal-derived biopolymers on the membrane surface and within the pores.

6.3 Theoretical background

6.3.1 Fouling in capillary membranes

The hydraulic performance of capillary MF/UF membranes is characterized by 3 main aspects,

- Fouling potential as MFI-UF;
- Pressure development during filtration;
- Build-up of resistance after backwash.

The Modified Fouling Index (MFI) was developed by Schippers and Verdouw (1980) as an index of the fouling potential of RO feed water for filtration in constant pressure based on cake filtration mechanism (See Chapter 3, section 3.2.1 for theoretical background). Further development by Boerlage et al. (2004) and Salinas Rodriguez (2011) resulted in the development of MFI-UF for constant flux filtration. In constant flux mode, MFI (or the fouling index, I) is determined from the slope of the linear region in a plot of ΔP versus time (See Chapter 5, section 5.2.1 for theory and formulation).

In this chapter, MFI-UF measured at constant flux through 150 kDa membranes (obtained from the first filtration cycle) was used as a tool to assess the fouling potential of AOM in synthetic seawater, in the presence and absence of coating. Assuming that cake/gel filtration is dominant in constant flux filtration, pressure development within successive cycles is given by,

$$\Delta P = \eta R_m J + \eta . I . J^2 . t \qquad \text{(Eq. 6-1)}$$

Where,
ΔP is applied pressure (bar)
η is viscosity (Pa.s)
R_m is clean membrane resistance (m^{-1})
J is filtration flux (L/m^2h)
I is fouling index (m^{-2})
t is filtration time (h).

Pressure (resistance to filtration) increases with filtration time (volume) until a certain threshold is reached at which membranes are cleaned by hydraulic backwashing. In most situations, backwashing is not effective in restoring permeability to its original level and pressure (resistance) increases after backwashing through successive filtration cycles (Figure 6-3a). Total resistance due to fouling R_t (m^{-1}) is the sum of backwashable (R_{BW}) and non-backwashable fouling resistance (R_{nBW}) and is given by,

$$R_t = R_{BW} + R_{nBW} = \frac{\Delta P}{\eta . J} \qquad \text{(Eq. 6-2)}$$

The total fouling rate, dR_t/dt (m^{-1}/h), is defined as the average slope of a linear regression of the last resistance values before backwash in successive filtration cycles as a function of time, while the average slope of a linear regression of the initial resistance values after backwash as a function of time is defined as the non-backwashable fouling rate dR_{nBW}/dt (m^{-1}/h); schematically presented in Figure 6-3a.

Deviation from linear pressure increase in time within filtration cycles might occur due to high fouling potential and compressibility of the cake/gel layer, enhanced compression of the cake/gel layer and/or enhanced rejection due to pore narrowing in the cake/gel layer. In this situation, average pressure increase per cycle (bar/h) is calculated (Figure 6-3b).

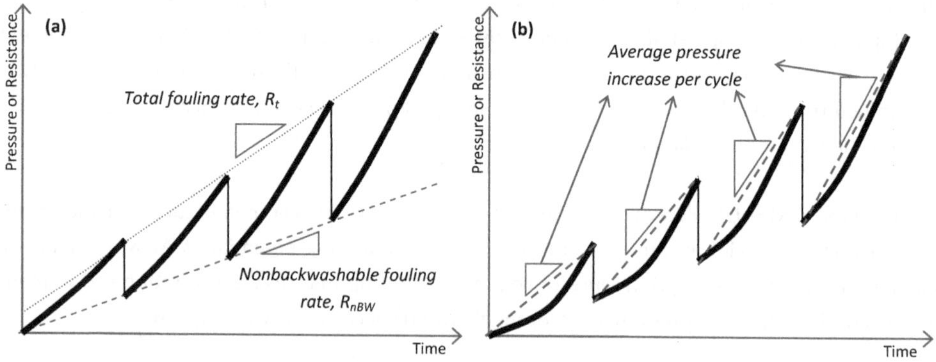

Figure 6-3 Determination of (a) total and non-backwashable fouling rates in constant flux UF operation and (b) average pressure increase per filtration cycle

6.3.2 Particle transport in capillary MF/UF

The success of coating is expected to depend on uniform coverage of the entire membrane surface by a layer of particles in a relatively short instance at the start of each filtration cycle. Coating material properties such as size and density affect particle transport through capillaries and may affect complete and uniform coverage of the membrane surface. Panglisch (2001) was the first to develop a model for the trajectory of spherical particles in dead-end filtration through capillary MF/UF membranes operated in inside-out mode. According to this model, a limiting particle diameter exists, beyond which lateral migration (tubular pinch effect) transports the particles away from the inlet and close to the dead-end of the capillary. The limiting diameter depends on operating and geometrical boundary conditions. The model was further adapted by Lerch (2008) to predict deposition of porous floc aggregates in capillary membranes operated in dead-end mode.

Several transport mechanisms may interact to bring particles close to the membrane or to preferentially transport them away from the membrane surface (Figure 6-4b). In MF/UF membranes, forces acting to bring particles toward the membrane include,

- Permeation drag;

- Van der Waals attraction between particles and membrane surface;
- Sedimentation;

while forces acting to transport particles away from the membrane (back-transport) include,

- Brownian diffusion;
- Shear induced diffusion;
- Buoyancy;
- Inertial lift (Green and Belfort, 1980; Altena et al., 1983; Belfort, 1989).

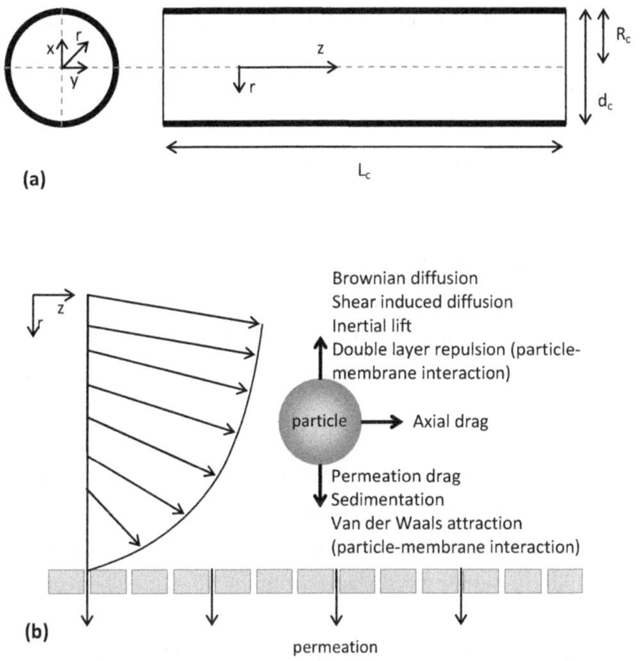

Figure 6-4 (a) Designation of coordinates inside a capillary membrane and (b) forces acting on a charged, spherical particle suspended in a viscous fluid under laminar flow conditions inside a capillary membrane (adapted from Panglisch, 2003)

Brownian diffusion

At steady state, particle accumulation at the membrane surface is balanced by diffusive back-transport to the bulk. Brownian diffusion can transport particles away from the membrane surface due to thermally induced random movement of particles.

Shear induced diffusion

Particle accumulation near the membrane surface can cause higher collision rates than in the bulk. These collisions can displace particles from flow streamlines and cause transportation of the particles away from the membrane surface. Shear induced diffusivity is highest near the membrane surface as particle collision is most frequent in areas with high particle density (Panglisch, 2001).

Inertial lift

Lateral migration, i.e., migration or lift of a particle perpendicular to the streamlines in the flow field, is driven by the fluid motion or its shear flow. Axial velocity is zero at the membrane surface and maximum in the centre of the capillary. Since the liquid at the top of the particle moves faster than that in the bottom, the particle spins clockwise. Particle spinning is most vigorous near the membrane surface due to the steepest velocity gradient. On the contrary, in the centre of the channel, particles do not spin at all, but move fastest along the streamline.

Sedimentation/buoyancy

Mechanisms of sedimentation and buoyancy are directly related to particle size and density. For neutrally buoyant particles - particles with a density close to that of water - the sedimentation and buoyancy forces may be neglected.

van der Waals attraction and double layer repulsion

These forces play a minor role in dilute suspensions and their effect may be neglected. Davis (1992) and Bowen et al. (1998) stated that particle-particle and particle-membrane interactions (including entropy, van der Waals interactions and electrostatic interaction) mainly play a role in concentrated solutions.

Based on the fluid flow field in a UF capillary, Panglisch (2003) calculated particle trajectories for different particle size and density by balancing the forces and torques acting on the particles in the flow field. Figure 6-5 depicts particle trajectories, where, r is particle distance from the centre of the capillary; r_c is capillary radius; z is distance from the fibre inlet; l_c is capillary length.

Particles entering at the centre of the capillary ($r/r_c = 0$) are subjected to the highest axial flow velocity and are transported to the dead end of the capillary, where they deposit. As r/r_c becomes larger, radial flow velocity has a more pronounced impact on the particles due to permeation drag strongly attracting the particles to the membrane surface. However, back-transport mechanisms occur that prevent particles from arriving to the membrane surface.

Calculations showed that neutrally buoyant particles (ϱ = 1,000 kg/m^3) with a diameter of 1 μm do not diverge from the streamlines in the flow field. However, for 15 μm particles of the same density, trajectories diverge substantially from the streamlines. These particles deposit at a significant distance from the capillary inlet. Lateral migration close to the membrane surface results in the transport of these particles for large distances along the capillary before deposition. Panglisch (2003) also showed that transport of non-neutrally buoyant particles (e.g., density 3000 kg/m^3) in a horizontal UF capillary is mainly governed by sedimentation.

The particle trajectory model is based on the assumption that the capillary membrane has uniform permeability; particle agglomeration before deposition can be neglected; particle suspension is mono-disperse; particles are distributed uniformly across the capillary entrance; once deposited, particles remain attached to the membrane surface.

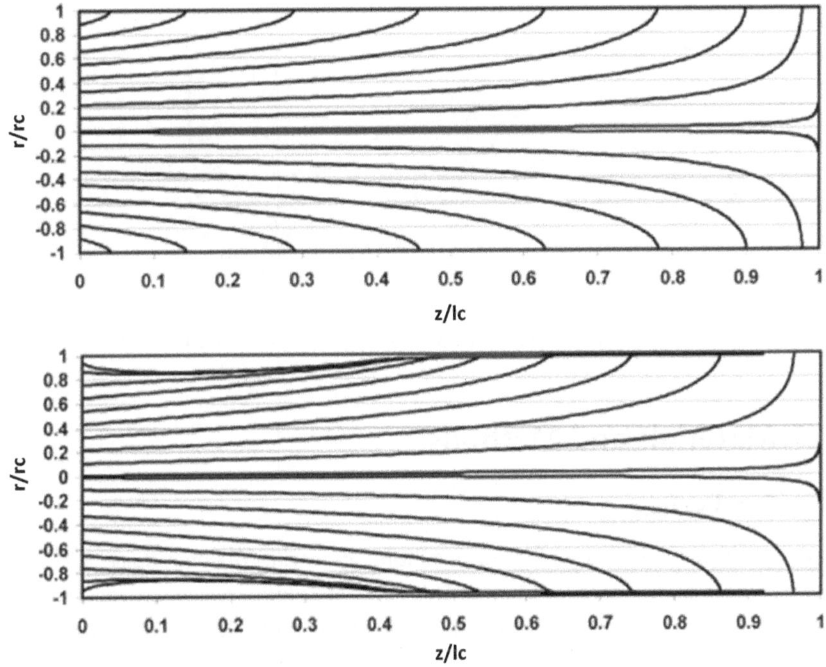

Figure 6-5 Particle trajectories for 1 μm particles (top) and 15 μm particles (bottom) entering the capillary along the radius; particle density 1,000 kg/m³, filtration flux 100 L/m²h

6.3.3 Particle deposition and membrane coverage

Panglisch (2003) determined theoretical deposition of spherical particles in a UF capillary by carrying out mass balances between the incoming and deposited particles. Π corresponds to percentage of particles with diameter d_p, which are deposited within an area with the length $0.1l_c$ (l_c is capillary length) to the amount of particles with diameter d_p entering the capillary.

Results obtained for particles of different size with a density of 1000 kg/m³, particle concentration of 100 mg/L, filtration flux of 100 L/m²h, capillary length of 1 m and capillary diameter of 0.7 mm are shown in Figure 6-6. Calculations show that smaller particles deposit in a wide, even distribution. Larger particles do not deposit until they are at a certain distance from the capillary inlet. The larger the particle size, the longer the distance. Panglisch (2003) validated these calculations by performing deposition experiments with spherical and non-spherical particles. Experimental results were in good accordance with theoretical calculations.

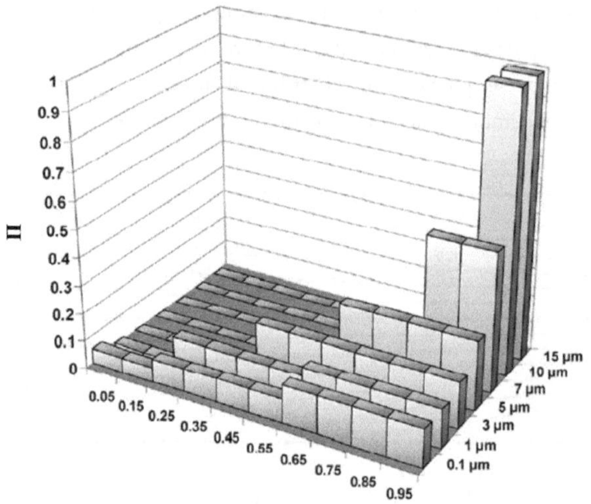

Figure 6-6 Particle deposition along the capillary length as a function of particle diameter (Panglisch, 2003)

6.4 Materials and Methods

6.4.1 Feed water

AOM was produced and isolated from a culture of marine diatoms, *Chaetoceros affinis*, in synthetic seawater as described in Chapter 5. The AOM solution was then analyzed by liquid chromatography-organic carbon detection (LC-OCD) to quantify the biopolymer fraction in the solutions. Feed water for the filtration experiments was prepared by diluting the AOM solution in synthetic seawater to reach a final biopolymer concentration of 0.2, 0.5 and 0.7 mg C/L.

6.4.2 Coating suspension

$FeCl_3.6H_2O$ analytical grade (Merck KGaA, Germany) was used to prepare coating suspensions. pH adjustment was done with 1M HCl and NaOH solutions. Two methods were used for preparing coating suspensions,

- Preliminary results (proof of principle) were performed with coating suspensions prepared at high concentration followed by low intensity grinding. Coating suspensions of 30 mg Fe(III)/L and 60 mg Fe(III)/L were prepared by precipitation of ferric hydroxide in synthetic seawater (TDS ~ 35 g/L) at pH 8 and gentle stirring for 15 minutes. Thereafter, grinding was applied at 1,100 s^{-1} for 1 minute.
- In the second method, coating suspensions were prepared by precipitation at low concentration followed by high intensity grinding. Coating suspensions were prepared at 3 mg Fe(III)/L and 10 mg Fe(III)/L in synthetic seawater (TDS ~ 35 g/L) at pH 8 and gentle stirring for 15 minutes. Thereafter, grinding at a G-value of 12,000 s^{-1} was applied for 1 minute. For comparison, grinding at G-values of 200 s^{-1}, 1,100 s^{-1} were also applied.

6.4.3 Characterization of coating particles

Particle size was measured with dynamic light scattering using a Zetasizer NanoZS (Malvern Instruments, UK). Measurements were done in triplicate for each sample. The Zetasizer NanoZS requires input data for absorbance of the suspension at λ = 632.8 nm and refractive index of the medium. Absorbance at 632.8 nm was measured with a UV-vis spectrophotometer and was approximately 0.007 and 0.021 for coating suspensions of 3 mg Fe(III)/L and 10 mg Fe(III)/L respectively. Refractive index of synthetic seawater was taken as 1.34 at 25 °C and salinity of 40 (Scripps Institution of Oceanography, USA).

A Philips CM120 transmission electron microscope (TEM) was used. After precipitation and grinding, a sample of approximately 8 µl was drawn from the coating suspension using a pipette and placed on a standard copper grid (3 mm in diameter). The water was allowed to drain (evaporate) before placing in the microscope.

6.4.4 Imaging

Images of the coated fibres were obtained using an Olympus BX51 light microscope with fitted digital camera (Olympus SC30, Japan). Scanning electron microscope (SEM) images were obtained with a JEOL JSM-35 CF equipped with imaging system. Prior to imaging, samples were mounted on a copper mount to ensure rapid thermal equilibration and sputtered with gold.

6.4.5 Particle deposition studies

To investigate coverage of UF membrane surface by coating particles, deposition of the coating suspension in capillary UF was studied. Experiments were performed with a single UF capillary of 0.75 m length, fed from one end only. This represents flow conditions in a full-scale module of 1.5 m fed from both ends, whereby the dead-end is at the centre of the fibre. Coating suspensions were prepared at three different grinding intensities of 200 s^{-1}, 1,100 s^{-1} and 12,000 s^{-1} and fed with a peristaltic pump (Cole Parmer, USA) to the module at 50 mg Fe(III)/m^2 at fluxes of 100 L/m^2h and 200 L/m^2h. Total iron arriving at the membrane was approximately 100 µg Fe(III). After coating, 6 segments of 5 cm length were cut from the inlet of the fibre to the dead-end. The segments were ultrasonicated for 30 minutes in 2% HNO_3 to bring all iron in solution. Iron measurement was performed with inductively coupled plasma - mass spectrometry (ICP-MS; Thermo Scientific XSeries 2) with a limit of detection of 0.002 mg Fe(III)/L. Based on measurements, iron concentration was calculated for 5 sections of 15 cm along the fibre length.

6.4.6 Filtration experiments

Experiments were performed using a bench-scale filtration setup schematically presented in Figure 6-7. Coating was applied at the start of each filtration cycle (at the same flux as filtration) followed by filtration of AOM solutions in synthetic seawater and backwashing. This procedure was repeated for multiple cycles. Coating suspensions were prepared fresh for each filtration experiment. The coating suspension was subjected to grinding for 1 minute prior to coating application in each cycle. Filtration was performed in constant flux dead-end mode at 100 L/m^2h for 30 minutes followed by backwashing at 250 L/m^2h for 1 minute. Coating was applied at 100 L/m^2h at equivalent dose ranging from 0.15 mg Fe(III)/L to 6 mg Fe(III)/L.

Modified polyethersulfone (PES) UF capillaries (Pentair X-flow, Netherlands) with a nominal MWCO of 150 kDa and inside diameter of 0.8 mm were used for filtration studies. Effective filtration area of the UF modules was 0.0045 m^2 ± 5%. UF pen modules were prepared in the laboratory by potting 6 UF capillaries of 30 cm effective filtration length in transparent flexible polyethylene tubing, PEN-x1, 25-NT (Festo, Germany) using polyurethane glue (Bison, The Netherlands).

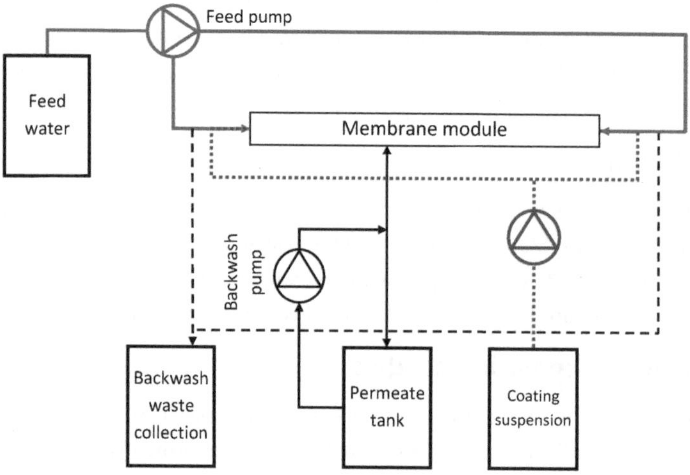

Figure 6-7 Scheme of laboratory-scale filtration setup for coating, filtration and backwash of capillary UF membranes

6.5 Results and Discussion

6.5.1 Fouling propensity of AOM solutions

Fouling propensity (pressure development and backwashability) of AOM was measured at different biopolymer concentrations (Figure 6-8). AOM fouling propensity strongly depended on biopolymer concentration in the feed solutions. For biopolymer concentration of 0.2 mg C/L, linear pressure development during filtration was observed, indicating that cake/gel filtration mechanism was dominant at this concentration of biopolymers. At higher biopolymer concentration, cake/gel filtration with compression was the dominant mechanism as pressure increase during filtration deviated substantially from linear curves.

In terms of fouling reversibility, non-backwashable fouling increased exponentially through successive cycles at higher biopolymer concentrations (Figure 6-9a). Average filtration slope per cycle (bar/h) followed a similar trend of exponential increase through successive cycles at higher biopolymer concentration (Figure 6-9b).

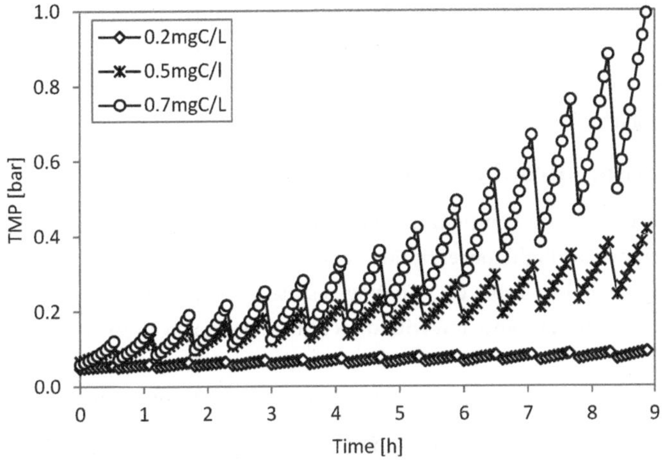

Figure 6-8 Filtration of AOM at different biopolymer concentrations at 100 L/m²h

A correlation was observed between average filtration slope per cycle and non-backwashable resistance through successive cycles. This correlation was particularly strong at higher biopolymer concentration. This indicated a reduction in effective filtration area due to complete blocking of the pores in certain areas resulting in higher local flux on the remaining available area for filtration (Refer Eq. 6-1).

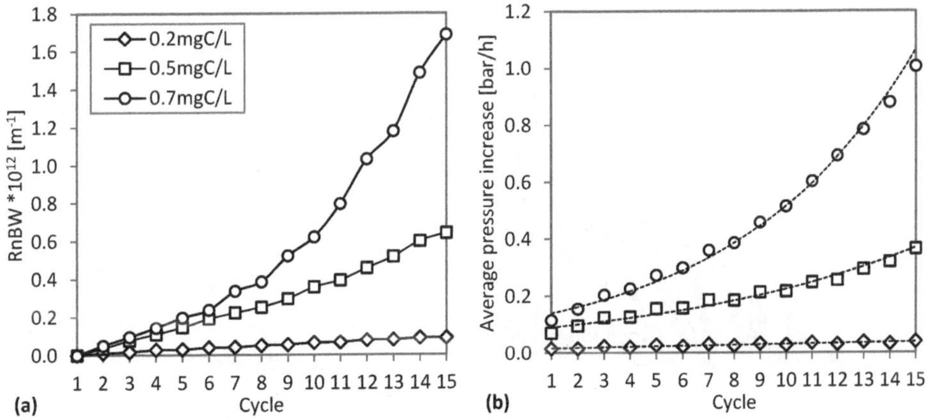

Figure 6-9 (a) Non-backwashable fouling resistance development and (b) average pressure increase per filtration cycle for AOM filtration at different biopolymer concentrations

An alternative explanation to pore blocking could be that backwashing was only effective in removing a portion of the cake/gel layer. AOM biopolymers are highly sticky and can attach strongly to the membrane surface. During backwashing a certain expansion and fluidization of the cake/gel layer may occur, followed by removal of the cake/gel layer by the cross flow (Marselina et al., 2011). If attractive forces are strong, this expansion might not take place or might be minimal. As a result only a small portion of the cake/gel layer is removed. The remaining cake/gel layer on the membrane surface is gradually more and more compressed through successive filtration cycles resulting in an exponential increase in pressure development and non-backwashable fouling resistance.

6.5.2 Coating - proof of principle

The following section is based on preliminary results obtained from coating UF membranes with suspensions prepared at high concentration followed by low intensity grinding. Experimental conditions are presented in Table 6-1. Coating resulted in lower overall pressure development for filtration during 6 hours compared with direct filtration (no coating) of AOM (Figure 6-10a).

Table 6-1 Experimental conditions for coating at different equivalent dose

Coating suspension conc. [mg Fe(III)/L]	Equivalent dose [mg Fe(III)/L]	Coating load * [mg Fe(III)/m^2]
	0.3	16
30	3	159
	6	318
	3	159
60	4	212
	6	318

* Calculated based on the assumption of uniform deposition.

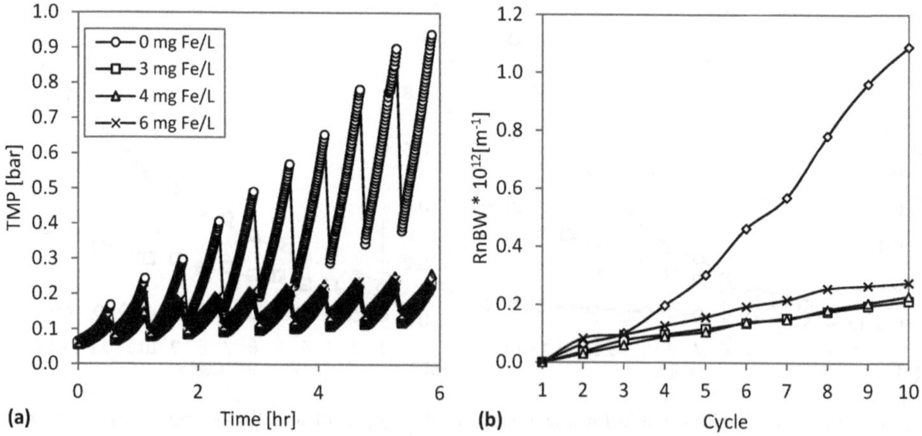

Figure 6-10 (a) Pressure development in time for filtration of AOM solutions (0.5 mg C/L as biopolymers) through UF membranes coated with particles created at 60 mg Fe(III)/L and grinding at 1,100 s^{-1}; (b) non-backwashable fouling resistance build-up

No difference was observed between dosing from coating suspensions of 30 and 60 mg Fe(III)/L. Development of non-backwashable resistance through successive filtration cycles is shown in Figure 6-10b. The coating layer applied at the start of each filtration cycle provides a protective physical barrier that prevents the strong attachment of sticky biopolymers to the membrane surface, resulting in a significantly lower rate of non-backwashable fouling. However, non-backwashable fouling of the coated membranes was not fully stabilized and an increase was observed through successive cycles for coating at all equivalent dose.

For filtration experiments performed with coating suspensions prepared at high concentration and low intensity grinding, relatively high equivalent dose was required. An attempt was made to lower the equivalent dose to 0.3 mg Fe(III)/L. Although non-backwashable fouling resistance was improved with coating at this low dose, operation could not be stabilized and resistance continued to increase with successive filtration cycles (Figure 6-11). This was attributed to non-uniform and incomplete coverage of the membrane surface by the coating layer.

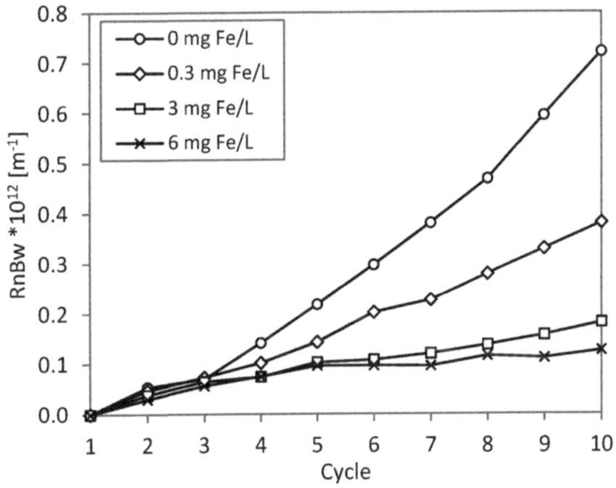

Figure 6-11 Non-backwashable fouling resistance as a function of coating equivalent dose for filtration of AOM at biopolymer concentration of 0.5 mg C/L; coating suspension of 30 mg Fe(III)/L

Autopsy of the UF capillaries revealed that most of the coating material was accumulated at the end of the fibres.

6.5.3 Characterization of coating particles

Theoretical calculations by Panglisch (2001; 2003) and Lerch (2008) showed that smaller particles deposit more evenly along the length of UF capillaries. As a consequence, it is expected that a more uniform coverage of the membrane surface can be obtained. To benefit optimally from uniform and full coverage, particles in the sub-micron range should be used. Tailoring particle size and morphology in sub-micron range is a complicated process (Mohapatra & Anand, 2010). As a starting point, a simple pathway based on precipitation in solution was chosen. In this pathway a large amount of iron (hydro-) oxide particles in the sub-micron range can be obtained.

In this process ferric hydroxide was precipitated in stock suspensions of low concentration (3 and 10 mg Fe(III)/L) at pH 8, followed by grinding at G-values of 200 s^{-1}, 1,100 s^{-1} and 12,000 s^{-1}.

Dynamic light scattering: For coating suspensions prepared by grinding at 200 s^{-1} and 1,100 s^{-1} particle size was outside the measurement range of the Zetasizer and could not be measured. As the measurement limit of the instrument is 10 μm, it can be assumed that particle size at these grinding intensities is larger than 10 μm. However this finding does not exclude the presence of smaller particles.

Particle size analysis of coating suspensions of 3 mg Fe(III)/L, measured immediately after grinding at 12,000 s^{-1}, showed particle size in the range of 200 nm to 1.5 μm, with an average particle size of approximately 540 ± 4 nm. For higher coating suspension concentration of 10 mg Fe(III)/L measured immediately after grinding, particle size was larger, in the range of 550 nm to 3 μm with an average particle size of approximately 990 ± 30 nm (Figure 6-12). Measurements were done in triplicate. For coating suspensions of 10 mg Fe(III)/L particles were visually larger prior to grinding (i.e., after precipitation) as compared with 3 mg Fe(III)/L.

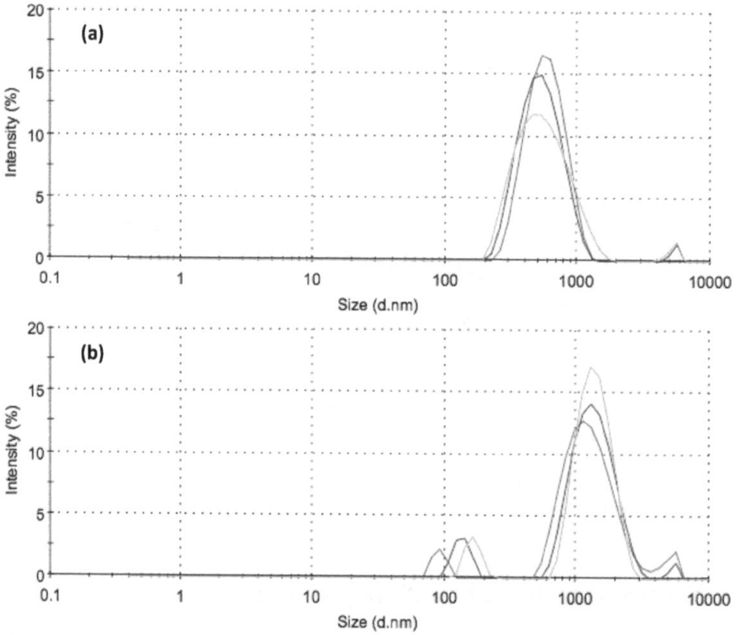

Figure 6-12 Particle size distribution of coating suspensions of (a) 3 mg Fe(III)/L and (b) 10 mg Fe(III)/L immediately after grinding at 12,000 s^{-1}

For measuring the size of particles/aggregates formed by precipitation of ferric hydroxide no particle size method can be considered ideal. Aggregates are highly irregular and porous and so their scattering patterns are likely to be very different than for equivalent solid spheres of the same material in light scattering devices. Farrow and Warren (1993) conclude that light/laser scattering and transmission techniques are useful for showing qualitative (rather than absolute) changes in floc size for aggregation systems.

Measurements of suspensions of 3 mg Fe(III)/L, 10 minutes after grinding, revealed larger particle size of approximately 705 ± 10 nm. For coating suspensions prepared at 10 mg Fe(III)/L, particle size was out of the measurement range of the instrument when measured 10 minutes after grinding. This finding has been taken into consideration in the filtration experiments, where coating procedure might take up to a few minutes for higher equivalent doses.

The effect of repeated grinding of the coating suspension on particle size was investigated. Particle size analysis showed that particle size did not change substantially with repeated grinding up to 7 times for coating suspensions of 3 and 10 mg Fe(III)/L. Based on this finding prior to coating application for each filtration cycle, grinding was applied to the suspension at the required G-values for 1 minute.

Transmission electron microscopy: TEM imaging was done to get an additional impression of the size and structure of these particles. Figure 6-13a shows TEM image of a coating particle isolated from a coating suspension of 3 mg Fe(III)/L prepared with grinding at 12,000 s^{-1}, with a particle size of approximately 600 nm. Electron diffraction (Figure 6-13b) of the same particle showed an amorphous structure.

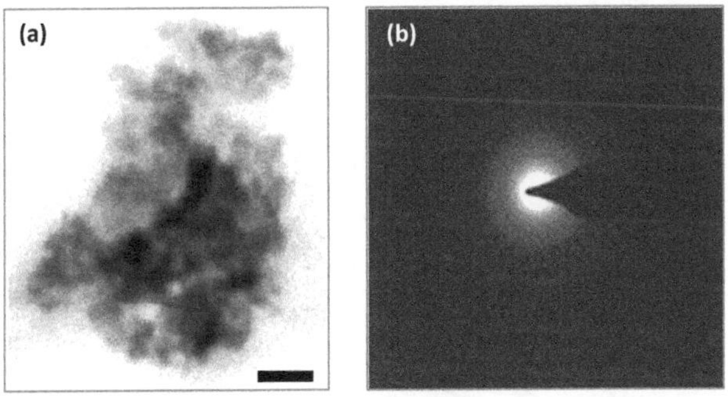

Figure 6-13 (a) TEM image of an isolated coating particle from stock coating suspension of 3 mg Fe(III)/L and grinding at 12,000 s^{-1}; scale bar 100 nm (b) electron diffraction of the same particle

Permeability and backwashability: Fouling potential and backwashability of the coating material itself was measured by coating the membranes before the start of each filtration cycle, followed by filtration of particle-free synthetic seawater for multiple cycles. Coating was applied at equivalent concentrations of 0.1, 0.15 and 0.3 mg Fe(III)/L at 100 L/m^{2}h. Thereafter synthetic seawater was filtered for 30 minutes through the coated membrane at 100 L/m^{2}h, followed by backwashing at 250 L/m^{2}h for 1 minute. Figure 6-14 shows resistance due to non-backwashable fouling for different equivalent dose. Contribution of the coating layers to increase in non-backwashable resistance was negligible.

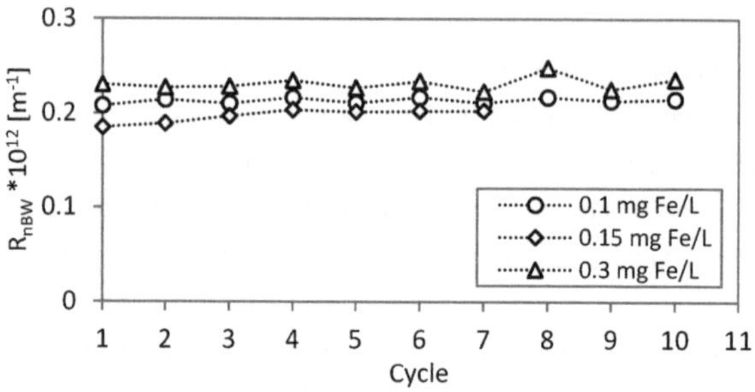

Figure 6-14 Non-backwashable fouling resistance build-up due to coating at different equivalent dose

6.5.4 Measured particle deposition

Deposition of iron hydroxide particles along the capillary with a length of 0.75 m, was determined by coating at 100 and 200 L/m^{2}h with suspensions of 10 mg Fe(III)/L. Coating suspensions were subjected to grinding intensities of 200 s^{-1}, 1,100 s^{-1} and 12,000 s^{-1} for a duration of 1 minute. Coating was done at 50 mg Fe(III)/m^2 (equivalent dose of 1 mg Fe(III)/L, considering filtration at 100 L/m^{2}h for 30 minutes), resulting in a total iron amount of approximately 100 μg Fe (III) arriving in the UF capillary. Figure 6-15 shows measured iron (III) amounts in five segments of the capillary normalized for total iron (III) dosed.

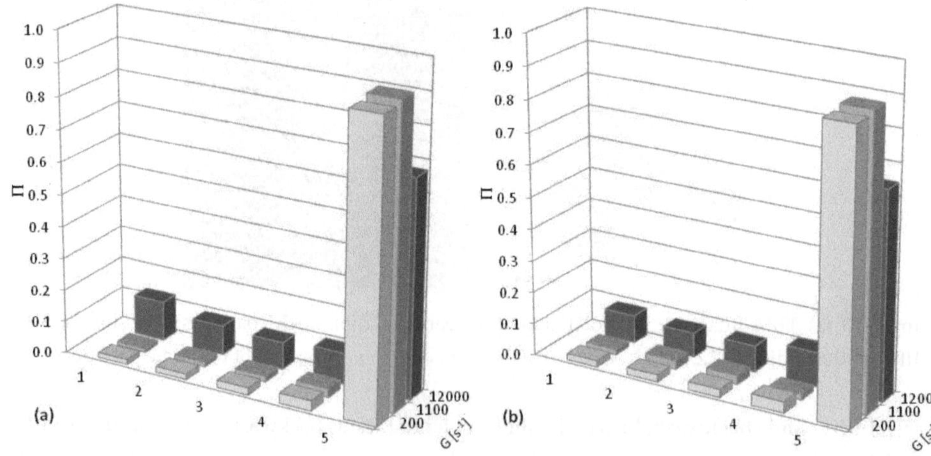

Figure 6-15 Iron deposition along a UF capillary as a function of grinding intensity (d_P) at (a) 100 L/m^{2}h and (b) 200 L/m^{2}h

For particles created at 12,000 s^{-1} (d_P 990 ± 30 nm), iron (III) deposited all along the capillary length. For this particle size, deposition was more or less uniform in the first four segments (0.6 m) of the capillary at about 10% per segment. Nonetheless, more than half of the particles deposited at the capillary dead-end. For particles prepared by grinding at 200 and 1,100 s^{-1} (average d_p > 10 μm) deposition was substantially less in the first four segments (0.6 m) of the capillary; between 1 and 4% per segment respectively. For these particle suspensions, iron (III) mainly accumulated in the last segment, at the capillary dead-end.

The effect of coating flux on deposition was marginal for particles prepared with three different grinding intensities, since no difference was observed between 100 and 200 L/m^{2}h. These observations are in concert with the findings of Panglisch (2003), further confirming that membrane coverage is higher and more uniform with smaller particles.

Cross-sections and longitudinal sections at various points along coated UF capillaries were observed visually, with a light microscope and a scanning electron microscope (Figure 6-16 and Figure 6-17). A relatively thick coating layer of up to a few micrometres was observed at the capillary dead-end (Figure 6-16c; Figure 6-17). Patches of uncovered membrane surface were observed in longitudinal sections (Figure 6-16b), indicating that the membrane surface was not fully covered by the coating layer.

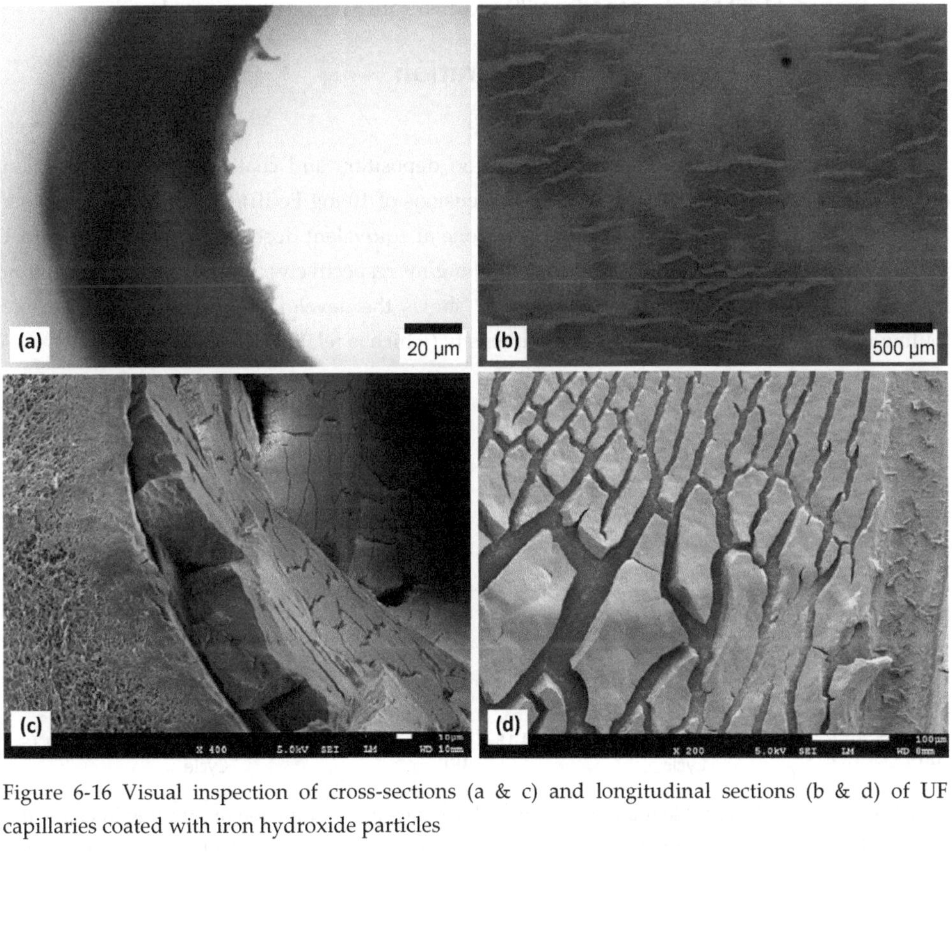

Figure 6-16 Visual inspection of cross-sections (a & c) and longitudinal sections (b & d) of UF capillaries coated with iron hydroxide particles

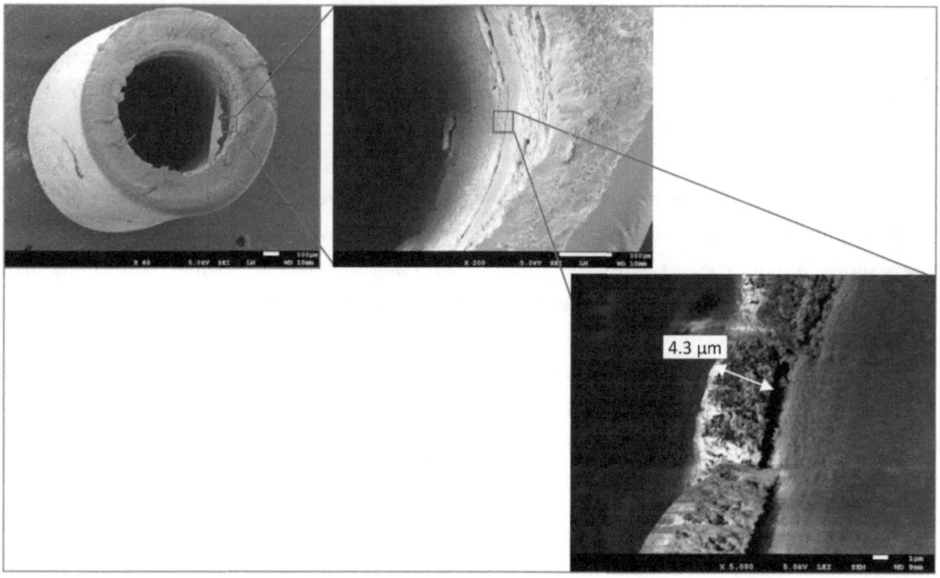

Figure 6-17 SEM images of a UF capillary coated with 3 mg Fe(III)/L equivalent dose

6.5.5 Effect of coating on UF operation

6.5.5.1 Particle size

To further illustrate the effect of particle size on deposition and coating efficiency, filtration experiments were carried out with coating suspensions of 10 mg Fe(III)/L prepared by grinding at 12,000 s^{-1}, 1,100 s^{-1} and 200 s^{-1}. Coating was done at equivalent dose of 0.5 mg Fe(III)/L and 1 mg Fe(III)/L (corresponding to 25 mg/m^2 and 50 mg/m^2 respectively). Biopolymer concentration in the feed water was 0.5 mg C/L. Figure 6-18 shows the development of non-backwashable fouling resistance as a function of grinding intensity (which is related to particle size).

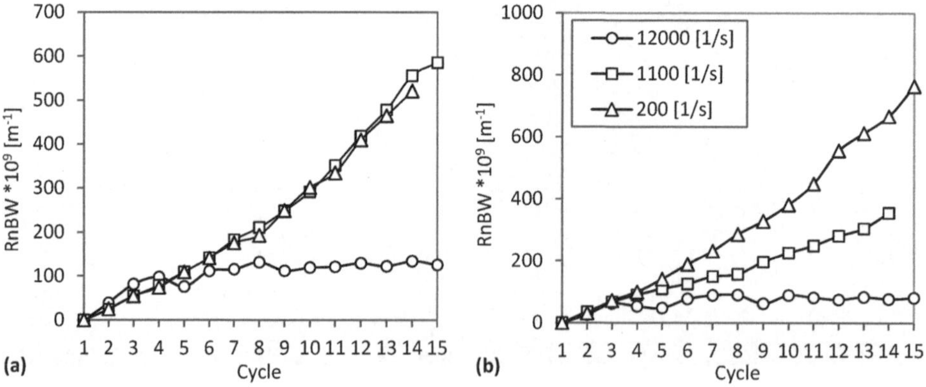

Figure 6-18 Non-backwashable fouling resistance development as a function of grinding intensity (G) at (a) equivalent dose of 0.5 mg Fe (III)/L (25 mg Fe(III)/m^2) and (b) 1 mg Fe (III)/L (50 mg Fe(III)/m^2)

The rate of non-backwashable fouling (Figure 6-19) was lower with coating than without coating. At higher grinding intensity (corresponding with smaller particle size) the rate of non-backwashable fouling development was substantially lower. When coating was performed with particles created at 12,000 s^{-1}, the difference in coating efficiency with 0.5 and 1.0 mg Fe(III)/L was marginal, indicating that full coverage of the membrane surface was achieved at 0.5 mg Fe(III)/L.

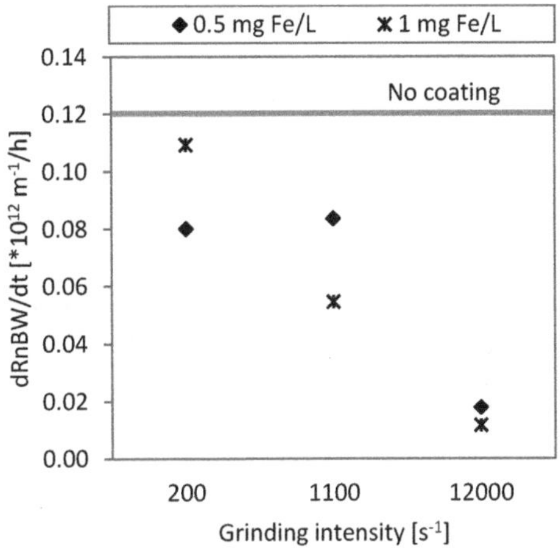

Figure 6-19 Effect of grinding intensity (G) on the rate of development of non-backwashable resistance at equivalent dose of 0.5 mg Fe (III)/L (25 mg Fe(III)/m^2) and 1 mg Fe (III)/L (50 mg Fe(III)/m^2)

Deposition efficiency of ferric hydroxide depended strongly on the grinding intensity at which the particles are formed. At grinding intensities of 200 s^{-1}, 1,100 s^{-1} and 12,000 s^{-1} this efficiency was approximately 1, 4 and 10 % respectively, indicating that particles smaller than 1 μm were most effective. Based on these results and observations, coating suspension concentration of 3 mg Fe(III)/L prepared by grinding at 12,000 s^{-1} was chosen for all subsequent experiments.

6.5.5.2 Effect of equivalent dose

To investigate the effectiveness of coating in more detail, coating was applied at different equivalent dose, i.e. from no coating to 1 mg Fe(III)/L equivalent dose. Equivalent dose and corresponding coating loads are presented in Table 6-2.

The development of non- backwashable fouling (resistance at the start of each filtration cycle) and the rate of increase of non-backwashable fouling are given in given in Figure 6-20 and Figure 6-21a respectively. Filtration in the absence of coating is marked by a steep increase in non-backwashable fouling resistance in subsequent cycles. Application of coating results in much lower non-backwashable fouling resistance for all doses. Coating at 0.15 mg Fe(III)/L and 0.3 mg Fe(III)/L equivalent dose, showed no improvement in non-backwashable fouling resistance for cycles 1-4. Operation improved from cycle 5 onwards, although a slight increase in resistance continued till cycle 12, after which a decline was observed.

Table 6-2 Coating loads and equivalent dose for experiments with coating suspension of 3 mg Fe(III)/L and grinding at 12,000 s^{-1}

Equivalent dose [mg Fe(III)/L]	Coating load [mg Fe(III)/m^2]
0.15	8
0.3	16
0.5	25
1.0	50
1.5	75

This indicates that for low equivalent dose, non-backwashable fouling development occurs to certain extent. This insufficient coating on the membrane surface, results in certain membrane area exposed to foulants. Once this exposed area is fouled, from cycle 5 onward, coating is sufficient to cover the remaining un-fouled membrane area and therefore stabilize operation. The decline in resistance from cycle 12 onwards may be due to the interference/attachment of coating particles to biopolymers that remain on the surface after backwashing.

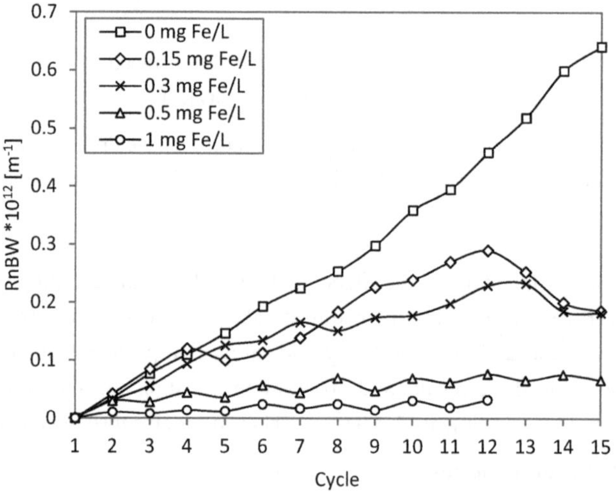

Figure 6-20 Effect of coating equivalent dose on non-backwashable fouling of UF membranes with AOM solutions of 0.5 mg C/L as biopolymers in synthetic seawater; coating suspensions prepared at 3 mg Fe(III)/L by grinding at 12,000 s^{-1}

Control of non-backwashable fouling was particularly notable for coating at 0.5 mg Fe(III)/L and 1 mg Fe(III)/L equivalent dose, where non-backwashable fouling was significantly reduced from cycle 2 onward and increased marginally through the filtration cycles. Non-backwashable fouling was marginal at an equivalent dose of 1.5 mg Fe (III)/L.

Figure 6-21b shows a remarkable effect of the coating on MFI-UF; the higher the dose the lower the MFI-UF. MFI-UF values are derived from the development of pressure during the first cycle for each coating experiment. This effect might be attributed to the occurrence of depth filtration in the coating, resulting in a slower - but linear - development of the resistance of the foulant layer (and pressure) in time. When coverage is more complete and the coating layer is thicker, the capacity of coating to facilitate depth filtration is larger, resulting in a lower MFI. This phenomenon is a beneficial side effect of coating.

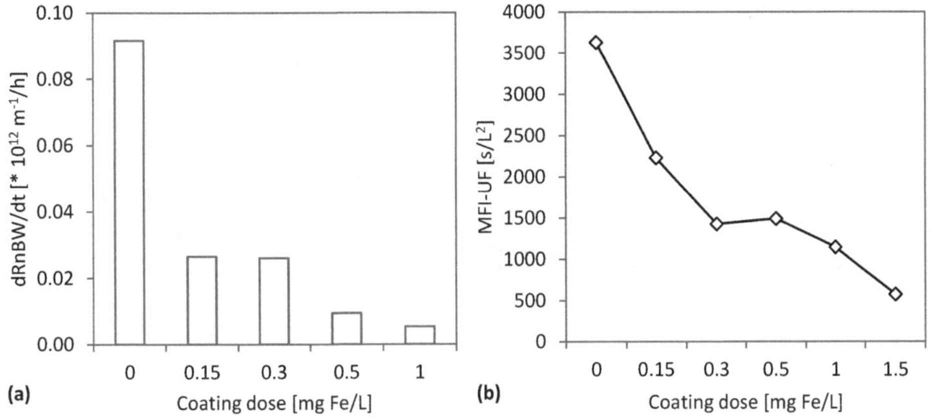

Figure 6-21 (a) Non-backwashable fouling rate (dR_{nBW}/dt) and (b) fouling potential (MFI-UF) as a function of coating dose at 100 L/m^2h

6.5.5.3 Biopolymer concentration

Experiments were conducted to investigate the effect of biopolymer concentration in feed water on required coating dose to control non-backwashable fouling in consecutive filtration cycles. For this purpose three biopolymer concentrations were tested; 0.2, 0.5 and 0.7 mg C/L for operation with no coating, 0.15 mg Fe(III)/L and 0.5 mg Fe(III)/L equivalent dose. Filtration flux was 100 L/m^2h. Development of non-backwashable fouling resistance during 15 consecutive filtration cycles is depicted in Figure 6-22.

A rapid development of non-backwashable resistance was observed at higher biopolymer concentrations when no coating was applied. Coating at equivalent dose of 0.15 mg Fe(III)/L resulted in a much slower development of non-backwashable resistance while coating at 0.5 mg Fe(III)/L gave a fast stabilization of the resistance at a low level. Figure 6-23 quantifies these observations in terms of non-backwashable fouling rate and fouling potential as measured by MFI-UF. Coating equivalent dose of 0.15 mg Fe(III)/L was evidently not sufficient to fully control non-backwashable fouling, indicating that at this dose full coverage of the membrane surface was not achieved.

A reason might be that the particles did not uniformly deposit on the membrane surface because an important part of the particles were too large to arrive at the membrane surface during coating. An additional reason might be that an important part of the particles arriving at the membrane surface do not attach to it. This last effect might explain why a coating of 0.15 mg Fe(III)/L becomes more effective after about 10 cycles in tests with 0.5 and 0.7 mg C/L, assuming that biopolymers present at the surface of the membrane will keep the coating material attached.

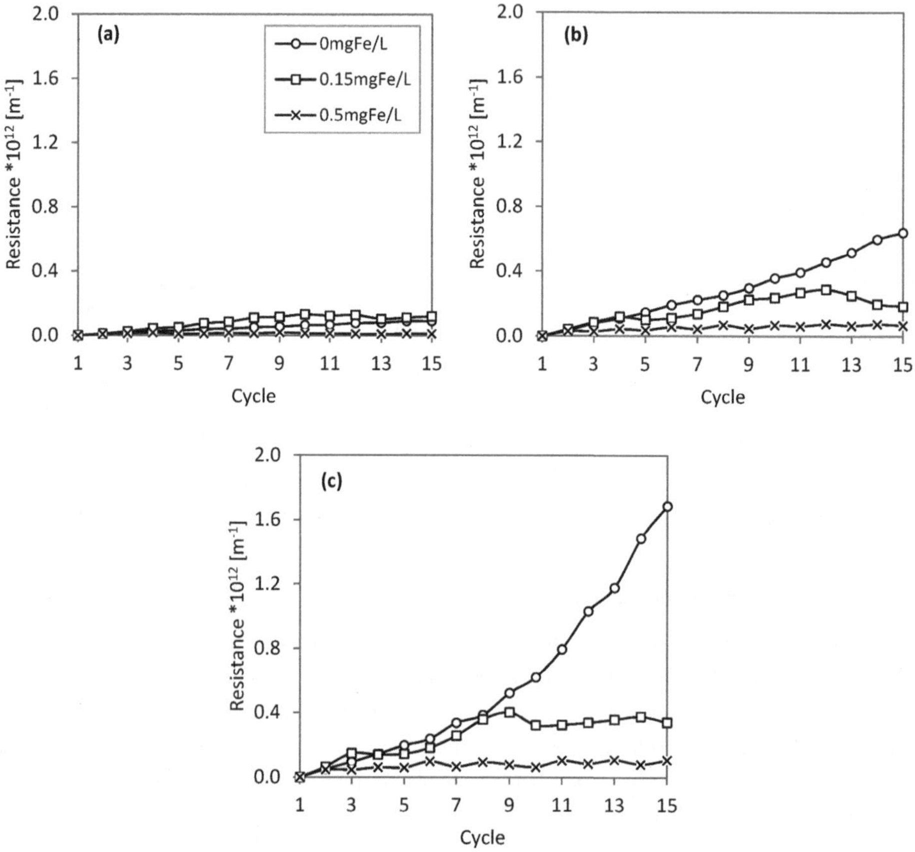

Figure 6-22 Non-backwashable fouling as a function of coating equivalent concentration for biopolymer concentrations of (a) 0.2 mg C/L, (b) 0.5 mg C/L and (c) 0.7 mg C/L

The effect of coating on fouling potential as measured by MFI-UF was particularly pronounced at coating equivalent dose of 0.5 mg Fe(III)/L for all biopolymer concentrations, whereas for coating with 0.15 mg Fe(III)/L this effect was only noticeable for biopolymer concentration of 0.7 mg Fe(III)/L.

These findings demonstrate that a very low coating dose is able to stabilize and control the non-backwashable fouling rate in presence of rather high biopolymer concentrations. Further improvements might be achieved, resulting in lower required dose, e.g., by making much smaller coating particles. However it is expected that additional research efforts will be required.

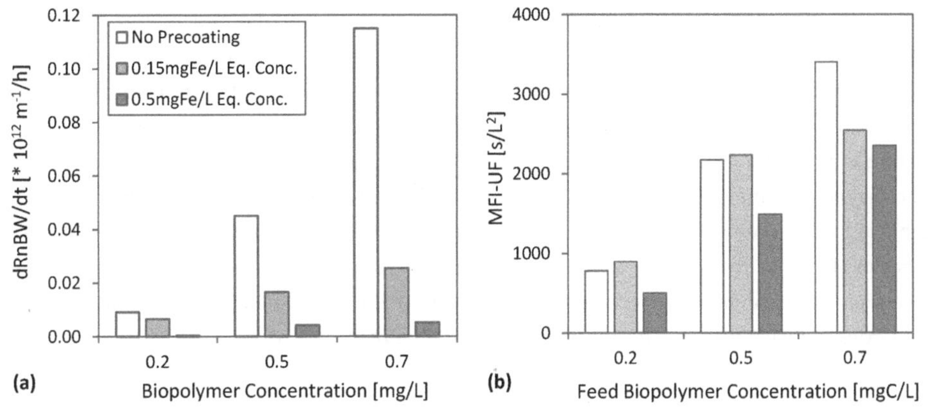

Figure 6-23 (a) Non-backwashable fouling rate and (b) MFI-UF as a function of biopolymer concentration for coated and non-coated UF membranes

6.5.5.4 Pre- and post-coating

Filtration tests were conducted in which ferric hydroxide coating was applied intermittently at approximately 14 cycles without and 14 cycles with coating (Figure 6-24). The membranes were first exposed to feed water without coating. Non-backwashable fouling resistance increased in this period steadily. From this point forward, coating was applied at 0.5 mg Fe(III)/L equivalent dose. Initially the non-backwashable resistance declined and then improved and stabilized. However, permeability was not yet fully restored. Coating was again switched off, and non-backwashable resistance increased rapidly during 14 cycles. Once again coating was applied at 0.5 mg Fe(III)/L equivalent dose. A very rapid decline and stabilization of non- backwashable resistance was observed. However, resistance could not be fully restored. Blankert et al. (2008) demonstrated a similar effect for inline coagulation with polyaluminium chloride (PACl) where resistance due to non-backwashable fouling was reversed by increasing coagulant dose.

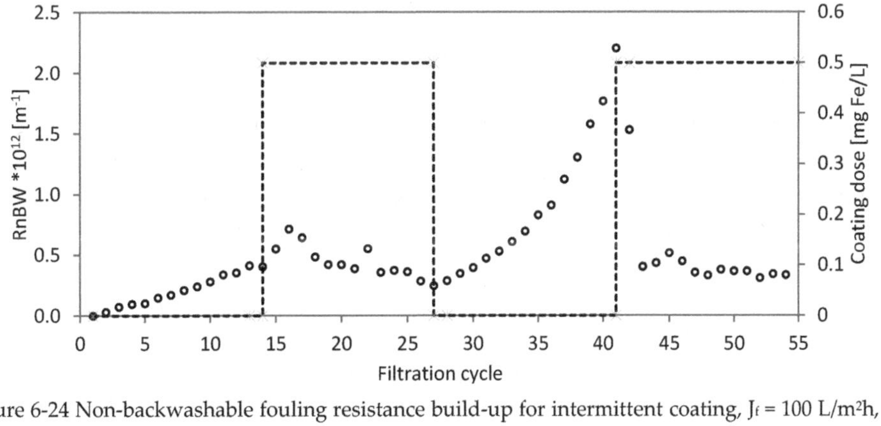

Figure 6-24 Non-backwashable fouling resistance build-up for intermittent coating, J_f = 100 L/m^{2}h, t_f = 30 min, J_{BW} = 250 L/m^{2}h, t_{BW} = 1 min

To test the effectiveness of coating, applied at the start and at the end of filtration cycles, experiments were conducted with 0.3 mg Fe(III)/L equivalent dose for a biopolymer concentration of 0.5 mg C/L. Filtration was performed at a flux of 100 L/m²h for 30 minutes, followed by backwashing at a flux of 250 L/m²h for 1 minute. Figure 6-25 shows that when coating is applied at the end of the filtration cycle, development of non-backwashable fouling is much slower compared to no coating. However, non-backwashable resistance is almost stabilized when coating is applied at the start of the filtration cycle, prior to feed water filtration. It is possible that the slightly positively charged ferric hydroxide particles interact with the negatively charged biopolymer particles attached to the membrane surface, changing their properties in such a way that backwashing is substantially improved.

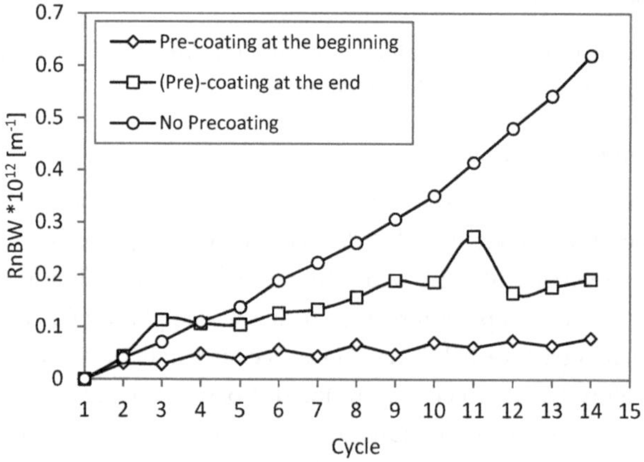

Figure 6-25 Coating application at the start versus the end of consecutive filtration cycles

This finding indicates that the mechanism with which coating improves UF operation may not be entirely physical, i.e., reducing contact area between the membrane surface and the sticky biopolymers, but (chemical) interactions between the coating material and foulants may play an important role as well.

6.5.5.5 Permeate quality

It was hypothesized that the pores of a coating layer of ferric hydroxide particles might be smaller than the pores of UF membranes, enhancing biopolymer rejection. For this purpose the effect of coating on permeate quality was investigated for 0.5 mg Fe(III)/L equivalent dose. Permeate of the first cycle was used for analysis with LC-OCD (Karlsruhe, Germany). Biopolymer concentration in the feed water was 0.5 mg C/L.

Figure 6-26 depicts that biopolymer concentration in permeate sample of the test with no coating is 10% higher, i.e., 220 µg C/L, than the biopolymer concentration in permeate of the coated membrane, i.e., 200 µg C/L. However, this small difference cannot be interpreted as significant, since it is in the range of reproducibility of the LC-OCD method. Limit of detection of the LC-

OCD method is 1-50 μg C/L depending on the fraction. This indicates that pre-coating with ferric hydroxide particles does not substantially improve permeate quality.

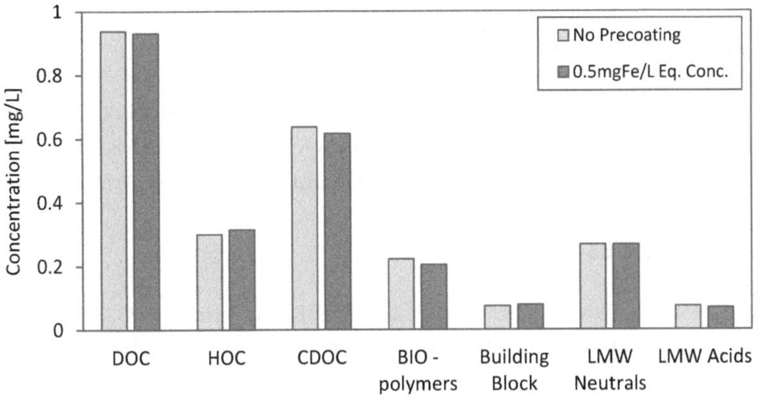

Figure 6-26 Permeate quality of UF membranes without coating and with coating at 0.5 mg Fe(III)/L equivalent dose; feed water biopolymer concentration 0.5 mg C/L

6.6 Conclusions and recommendation

Biopolymers originating from the marine diatom *Chaetoceros affinis* showed a high fouling potential measured as MFI-UF and demonstrated a poor backwashability in filtration tests with 150 kDa UF membranes.

Application of a coating layer - consisting of ferric hydroxide prepared by precipitation at high concentration followed by low intensity grinding - at a dose of 3 to 6 mg Fe(III)/L at the start of each filtration cycle resulted in stabilizing and controlling the rate of non-backwashable fouling.

Pre-coating with ferric hydroxide particles in the range of 0.5 to 1 μm - produced by precipitation and high intensity grinding - enabled to stabilize and control the rate of non-backwashable fouling at very low equivalent dose of ≤ 0.5 mg Fe(III)/L for filtration of solutions containing 0.2 to 0.7 mg C/L biopolymers.

The success of coating in reducing non-backwashable fouling of UF membranes is attributed to a combination of physical and chemical mechanisms. Results indicated that physical shielding may not be the only mechanism - as initially hypothesized - and (chemical) interaction between coating material and the biopolymers attached to the membrane surface also occurs in such a way that non-backwashable fouling is reduced.

Further reduction of the coating dose might be possible by e.g., applying smaller particles - since uniform layers were not formed on the membranes surface - and is recommended for further research.

To be able to distinguish between physical and chemical interactions between coating layer and foulants, it is recommended to reproduce the experiments with inert particles (e.g., polystyrene nanoparticles).

6.7 Acknowledgements

Thanks are due to Anil Ozan from the Department of Applied Sciences of TU Delft for TEM imaging, Dr. Stefan Huber from DOC-Labor for LC-OCD analyses, Dr. Bastiaan Blankert for insightful discussions, Abubakar Hassan, Bayardo Gonzalez and Clement Trellu.

6.8 References

Altena, F.W., Belfort, G., Otis, J., Fiessinger, F., Rovel, J.M., Nicoletti, J., 1983. Particle motion in a laminar slit flow: A fundamental fouling study. Desalination 47 (1-3), 221-232.

Belfort, G., 1989. Fluid mechanics in membrane filtration: recent developments. Journal of Membrane Science 40 (2), 123-147.

Blankert, B., Betlem, B.H.L, Roffel, B., 2007. Development of a control system for in-line coagulation in an ultrafiltration process. Journal of Membrane Science. 301, 39-45.

Boerlage, S.F.E., Kennedy, M., Tarawneh, Z. Faber, R.D., Schippers, J.C., 2004. Development of the MFI-UF in constant flux filtration. Desalination 161, 103–113.

Bowen, W.R., Hilal, N., Lovitt, R.W., Wright, C.J., 1998. A new technique for membrane characterisation: Direct measurement of the force of adhesion of a single particle using an atomic force microscope. Journal of Membrane Science 139, 269-274.

Brink, L.E.S., Romijn, D.J., 1990. Reducing the protein fouling of polysulfone surfaces and polysulfone ultrafiltration membranes. Optimization of the type of presorbed layer, Desalination 78, 209.

Caron , D.A., Garneau, M.E., Seubert, E., Howard M.D.A., Darjany L., Schnetzer A., Cetinic, I., Filteau, G., Lauri, P., Jones, B., Trussell, S., 2010. Harmful algae and their potential impacts on desalination operations off southern California. Water Research 44 (2), 385-416.

Davis, R.H., 1992. Modelling of fouling of crossflow microfiltration membranes. Separation and Purification Methods 21 (2), 75-126.

Fogg, G.E., 1983. The ecological significance of extracellular products of phytoplankton photosynthesis. Bot. Mar. 26, 3–14.

Galjaard, G., Buijs, P., Beerendonk, E., Schoonenberg, F., Schippers, J.C. 2001a. Precoating (EPCE®) UF membranes for direct treatment of surface water. Desalination 139, 305-316.

Galjaard, G., van Paassen, J., Buijs, P., Schoonenberg, F., 2001b. Enhanced pre-coat engineering (EPCE) for micro- and ultrafiltration: the solution for fouling? Water Science and Technology: Water Supply 1(5/6), 151-156.

Galjaard, G., Kruithof, J.C., Scheerman, H., Verdouw, J., Schippers, J.C., 2003. Enhanced pre-coat engineering (EPCE®) for micro- and ultrafiltration: steps to full-scale application. Water Science and Technology: Water Supply 3(5-6), 125-132.

Green, G., Belfort, G., 1980. Fouling of ultrafiltration membranes - Lateral migration and the particle trajectory model. Desalination 35, 129-147.

Henderson, R.K., Baker, A., Parsons, S.A., Jefferson, B., 2008. Characterisation of algogenic organic matter extracted from cyanobacteria, green algae and diatoms. Water Research 42, 3435-3445.

Khayet, M., Suk, D.E., Narbaitz, R.M., Santerre, J.P., Matsuura, T., 2003. Study on surface modification by surface-modifying macromolecules and its applications in membrane-separation processes, Journal of Applied Polymer Science 89, 2902.

Kim, K.J., Fane, A.G., Fell, C.J.D, 1988. Performance of ultrafiltration membranes pretreated by polymers, Desalination 70, 229.

La, Y-H., McCloskey, B.D., Sooriyakumaran, R., Vora, A., Freeman, B., Nassar, M., Hedrick, J., Nelson, A., Allen, R. 2011. Bifunctional hydrogel coatings for water purification membranes: Improved fouling resistance and antimicrobial activity. Journal of Membrane Science 372, 285-291.

Lerch, A. 2008. Fouling layer formation by flocs in inside-out driven capillary ultrafiltration membranes. Dissertation, Universität Duisburg-Essen.

Marselina, Y., Le-Clech, P., Stuetz, R.M., Chen, V., 2011. Characterisation of membrane fouling deposition and removal by direct observation technique. Journal of Membrane Science 341, 163-171.

Myklestad, S., Haug, A., Larsen, B., 1972. Production of carbohydrates by the marine diatom Chaetoceros affinis Var. Willei (Gran) Hustedt. II. Preliminary Investigation of the extracellular polysaccharide. Journal of Exp. Mar. Biol. Ecol. 9, 137-144.

Myklestad, S., 1995. Release of extracellular products by phytoplankton with special emphasis on polysaccharides. Science of the Total Environment 165, 155-164.

Panglisch, S. 2001. Zur Bildung und Vermeidung schwer entfernbarer Partikelablagerungen in Kapillarmembranen bei der Dead-End Filtration, Dissertation, IWW-Mülheim.

Panglisch, S., 2003. Formation and prevention of hardly removable particle layers in inside-out capillary membranes operating in dead-end mode. Water Science and Technology: Water Supply 3 (5-6), 117-124.

Passow, U., Alldredge, A. L., & Logan, B. E. (1994). The role of particulate carbohydrate exudates in the flocculation of diatom blooms. Deep-Sea Research I, 41, 335–357.

Passow, U., Alldredge, A.L., 1994. Distribution, size, and bacterial colonization of transparent exopolymer particles (TEP) in the ocean. Marine Ecology Progress Series 113, 185-198.

Salinas Rodriguez, S.G., 2011. Particulate and organic matter fouling of SWRO systems: Characterization, modelling and applications. PhD dissertation, UNESCO-IHE/TU Delft, Delft.

Schippers, J.C., Verdouw, J., 1980. The modified fouling index, a method of determining the fouling characteristics of water. Desalination 32, 137-148.

Schurer R., Tabatabai A., Villacorte L., Schippers J.C., Kennedy M.D., 2013. Three years operational experience with ultrafiltration as SWRO pre-treatment during algal bloom. Desalination and Water Treatment 51(4-6), 1034-1042.

Susanto, H., Ulbricht, M. 2006. Performance of surface modified polyethersulfone membranes for ultrafiltration of aquatic humic substances. Desalination. 199, 384-386.

Ulbricht, M., Belfort, G., 1996. Surface modification of ultrafiltration membranes by low temperature plasma II. Graft polymerization onto polyacrylonitrile and polysulfone, Journal of Membrane Science 111, 193–215.

Watt, W. D. (1969). Extracellular release of organic matter from two freshwater diatoms. Annual Botany, 33, 427–437.

Zhou, J., Mopper, K., & Passow, U., 1998. The role of surface-active carbohydrates in the formation of transparent exopolymer particles by bubble adsorption of seawater. Limnology and Oceanography. 43, 1860–1871.

7

REMOVAL OF ALGAL ORGANIC MATTER FROM SEAWATER WITH COAGULATION

There is growing evidence that algal organic matter (AOM) released copiously during periods of algal bloom can cause serious particulate and biofouling in reverse osmosis (RO) membranes, resulting in frequent need for membrane cleaning. Consequently removal of this type of organic matter through effective pretreatment is highly desirable. This chapter investigated the effect of coagulation on AOM removal from seawater as a function of coagulant dose at pH 5 and 8. Synthetic seawater spiked with AOM obtained from a marine diatom, *Chaetoceros affinis* (cultivated in the laboratory) was treated by conventional coagulation with ferric chloride followed by sedimentation and filtration through 0.45 μm. Coagulation followed by sedimentation was effective in reducing biopolymers concentration (as measured by LC-OCD) and fouling potential (as measured by MFI-UF) by up to 70% and 80% respectively. Coagulation followed by sedimentation was also effective in removing $TEP_{0.4}$ by more than 80% for coagulant dose > 5 mg Fe(III)/L. Biopolymers and $TEP_{0.4}$ removal was not significantly different for coagulation at pH 5 and pH 8. Intensity peaks observed from F-EEM analysis confirmed the assumption that most of the AOM is comprised of polysaccharide-like material, as the protein peaks were not significant. Filtration through 0.45 μm had a pronounced impact on AOM removal for all parameters analyzed, even at coagulant dose < 1 mg Fe(III)/L indicating that flocs formed from coagulation of AOM have better filterability than settlability characteristics.

7.1 Introduction

Algae are prominent in fresh and saline surface water bodies and play a significant role in aquatic ecology. In marine environments, phytoplankton provide the base for most food chains. Seasonal population explosion of algae - algal bloom - results in high density of algal cells, release of cell fragments and dissolved or colloidal organic matter into water bodies. This can pose significant problems to aquatic environments by causing discolouration, toxicity, mortality of marine organisms and odour. Algal blooms are caused by various bloom-forming phytoplankton species, some of which commonly give rise to green tides (e.g., microcystis), brown tides (e.g., diatoms) and red tides (e.g., dinoflagellates).

For drinking water production from fresh water, algae and their associated organic matter have been reported to adversely affect treatment processes by causing rapid clogging of media filters and breakthrough of algae (Bernhardt and Clasen, 1991). From the perspective of seawater reverse osmosis (SWRO) plants, algal blooms can adversely affect the operation of both pretreatment systems and reverse osmosis (RO) membranes. The red tide event in the Gulf of Oman in 2008-09 that resulted in the shutdown of several RO desalination plants in the region and caused substantial fish kill marked a major point of concern for the RO desalination industry. All SWRO plants that reported shut down in the region are equipped with conventional pretreatment (i.e., dual media filtration). Algal organic matter (AOM) released during algal blooms can cause significant deterioration in performance of dual media filters (DMF) due to severe clogging of the filters resulting in short run lengths and reduced production capacity. This, together with concerns regarding RO feed water quality (SDI > 5) as a result of breakthrough of particles, mostly likely AOM, caused the shutdown of plants in the region.

AOM is composed of intracellular organic matter (IOM) formed due to autolysis consisting of proteins, nucleic acids, lipids and small molecules and extracellular organic matter (EOM) formed via metabolic excretion and composed mainly of acidic polysaccharides (Fogg, 1983). A significant fraction of AOM is composed of high molecular weight, hydrophilic, anionic extracellular polysaccharides and glycoproteins. These substances are present in both fresh and seawater, and are known as transparent exopolymer particles (TEP). TEP were introduced by Alldredge et al. (1993) in seawater and are extensively studied in oceanography as they play a major role in the aggregation dynamics of algae during blooms (Myklestad, 1995; Mopper et al., 1995).

Berman and Holenberg (2005) were the first to suggest that TEP could potentially enhance biofouling in RO membranes by providing a conditioning layer for bacterial deposition, attachment and growth. Once TEP adhere to the membrane surface, they can provide a nutritious substrate for microbial growth and establishment of biofilm. Therefore, removal of particulate and colloidal TEP and their precursors through effective pretreatment is desired.

Most SWRO desalination plants are equipped with conventional pretreatment systems, namely, dual media filtration (DMF) and advanced pretreatment systems namely micro- or ultrafiltration (MF/UF). Conventional systems frequently rely on coagulant addition to meet RO feed water quality requirements, which is mainly focused on turbidity and SDI. A growing share of SWRO systems incorporates MF/UF as pretreatment. Some of these systems apply coagulants as well to control the rate of fouling of MF/UF membranes during algal blooms. AOM removal from RO feed water by pretreatment is not commonly determined and very limited data is available.

Analytical parameters that are commonly applied for the characterization of natural organic matter (NOM) in water and wastewater treatment processes are conventional surrogate parameters,

- TOC/DOC for the measurement of bulk organic matter in water samples,
- UV_{254} for specifying the presence of aromatic organic compounds,
- Specific UV absorbance (SUVA) - normalized UV_{254} absorbance with respect to overall organic load.

TOC/DOC only reflects the bulk of organic matter and does not give any information on the nature and composition of organic matter or the contribution of different fractions to the overall concentration. UV_{254} and SUVA may not be adequate since Henderson et al. (2008) observed that AOM obtained from four algal species (including a marine diatom) were dominated by hydrophilic polysaccharides with low SUVA values, signifying a lack of UV_{254} absorbing compounds. Furthermore, as the name suggests, TEP are transparent, therefore turbidity measurements cannot provide information on their presence or removal rates. Advanced techniques for organic matter characterization might give more relevant information. These techniques include,

- Liquid chromatography - organic carbon detection (LC-OCD) which provides information on the nature and concentrations of different fractions of NOM based on their molecular weight (Huber et al., 2011),
- Fluorescence excitation emission matrices (F-EEM) which can provide specific information on fluorescent fractions of organic matter (Her et al., 2004).

Modification and improvement of TEP measurement methods by Villacorte et al. (2013) have provided further insight on the presence and concentration of these particles in seawater and along various treatment schemes.

In practice, coagulation efficiency in SWRO pretreatment is quantified in terms of SDI, turbidity and residual coagulant. However meeting the guidelines for these parameters does not guarantee low fouling rate of the RO membranes. This is primarily attributed to the inadequacy of these parameters in predicting particulate/organic and bio-fouling. To improve prediction of these types of fouling, several new parameters have been developed and several others are under development.

Examples of such parameters are,

- Modified fouling index-ultrafiltration (MFI-UF) for particulate fouling (Salinas Rodriguez, 2011; Boerlage et al., 2004; Schippers and Verdouw, 1980);
- TEP measurement (Passow and Alldredge, 1995; Villacorte et al., 2013);
- Measurement of biopolymers concentration by LC-OCD (Huber et al., 2011);
- Membrane fouling simulator tests to quantify biofouling potential (Vrouwenvelder, 2007) in addition to existing methods for measuring assimilable organic carbon (van der Kooij, 1982) and biodegradable organic carbon (Joret et al., 1988; Volk et al., 1994).

This chapter focused on coagulation of AOM in synthetic seawater and the effect of coagulant dose and pH on treated water quality. Solutions of AOM in synthetic seawater are coagulated in a jar test apparatus for different coagulant concentrations at pH 5 and 8, for fixed mixing, flocculation and settling, followed by 0.45 μm filtration. Coagulation efficiency for settled and filtered water was investigated in terms of surrogate parameters such as TOC/DOC, UV_{254}, SUVA; advanced parameters related to AOM such as LC-OCD (for biopolymers detection), F-EEM, $TEP_{0.4}$; and particulate fouling related parameters such as $MFI\text{-}UF_{10kDa}$.

7.2 Materials and methods

Algal blooms are seasonal phenomena and their occurrence is hard to predict. To ensure feed water availability of certain quality criteria, a strain of marine diatoms were cultivated in the laboratory to produce AOM. It is not unlikely that AOM produced from different algal species may exhibit different properties. In this study we aimed at measuring removal rates of at least 90% for various surrogate parameters, which set requirements for the minimum level of AOM in feed water and for the level of detection (LoD) of the surrogate parameters used.

7.2.1 Algae cultivation

A pure strain of *Chaetoceros affinis* (CCAP 1010/27) was inoculated in sterilised synthetic seawater (SSW). *Chaetoceros affinis* belongs to a large genus of marine planktonic diatom species that often form blooms in the North Atlantic and are among the species known to produce the highest amounts of TEP in seawater (Passow, 2002). Cultures were grown at room temperature in f/2+Si medium (Guillard and Ryther, 1962) for marine diatoms in 2 L flasks with continuous shaking at moderate speed to ensure nutrient distribution and avoid settling of cells. Cultures were exposed to a 12/12h light/dark cycle using mercury fluorescent lamps with an incident photon flux density of 40-50 $\mu mol.m^{-2}.s^{-1}$.

7.2.2 AOM production

AOM was harvested from the algal cultures during stationary growth phase, approximately 14 days after inoculation. The flasks were removed from the shakers and allowed to settle for at least 24 hours. Supernatant was separated by suction without disturbing the settled algae at the bottom of the flasks. Samples were not filtered to avoid loss of AOM from the solutions. From a practical point of view this represents natural AOM release during algal blooms. The AOM is assumed (not verified) to be mostly extracellular algal organic matter (EOM), as no cell destruction was attempted.

Inoculation and harvesting of AOM was performed in batches, limited in size by laboratory capacity in terms of glassware, lamps and shakers. Potential differences in coagulation behaviour of different batches due to variations in AOM characteristics produced per batch cannot be excluded but will be investigated in the following sections. In this study two distinct batches of AOM were prepared for the experiments as follows,

- Batch 1, biopolymers concentration 2.33 mg C/L. Experiments at pH 8 (all analyses)
- Batch 1, biopolymers concentration 2.33 mg C/L. Experiments at pH 5 (all analyses except $TEP_{0.4}$)
- Batch 2, biopolymers concentration 2.52 mg C/L. Experiments at pH 5 (only $TEP_{0.4}$)

7.2.3 Feed water

In order to provide the desired salinity matrix and to reduce interference from (in)organic contaminants in real seawater, synthetic seawater (SSW) was prepared based on the typical ion concentration of coastal North Sea water (TDS ~ 35 g/L). Analytical grade salts were sequentially dissolved in Milli-Q water. Bromide, fluoride, boron and strontium were not considered in the SSW matrix, as their concentrations are typically very small in comparison to other inorganic ions. Batches of 9 litres were prepared in a 10 litre round bottom flask, stirred with magnetic stirrer for at least 24 hours and kept at room temperature (~20 °C) until used. Measured pH and conductivity of SSW were 8.0 and 49,000 μS/cm respectively. Conductivity and pH of SSW were used to verify the quality of different batches.

Feed water was prepared by fixing the initial biopolymers concentration to 0.5 mg C/L as measured by LC-OCD. This is the highest concentration measured in North Sea water during algal bloom (2011) conditions in the period 2009-2012. AOM stock solutions (biopolymer concentration = 2.33 and 2.52 mg C/L for Batch 1 and 2 respectively) were diluted in SSW with a TDS of 35 g/L to obtain a final concentration of 0.5 mg C/L.

7.2.4 Coagulation, settling and filtration

The effect of coagulation with ferric chloride on AOM removal was investigated based on organic matter surrogate parameters such as TOC/DOC, UV_{254}, SUVA; LC-OCD (for biopolymers detection), F-EEM and $TEP_{0.4}$; and MFI-UF.

7.2.4.1 Coagulation experiments

Batch experiments were performed for different coagulant dose at pH 5 and 8. Experiments were performed with an EC Engineering Jar Tester, Model CLM4 (Alberta, Canada). Ferric chloride (PIX-111, KEMIRA) was added to 1 L solutions of AOM in SSW at the start of a 20 sec rapid mixing step (1,100 s^{-1}). Prior to coagulant addition, pH was adjusted to obtain a pH of 5 or 8 for coagulation by adding 1M HCl or NaOH as required. Even though H_2SO_4 is commonly used to adjust feed water pH, we used HCl in our experiments as sulphuric acid addition has the potential to cause sulphate precipitates (Greenlee et al., 2009).

Flocculation was performed for 15 min by slow mixing at 45 s^{-1}, after which the paddles were gently removed and the solutions were allowed to settle for 20 min. Settled water samples were collected from sampling ports located 10 cm from the base of the jar. Filtered samples were prepared by filtering settled water samples through 0.45 μm cellulose acetate filters (Millipore, USA) at a filtration rate of 10 mL/min. Studies on coagulation of (in)organic particles in surface waters frequently use 0.45 μm filtration to simulate the performance of media filtration on water quality. Experimental conditions are tabulated in Table 7-1.

Table 7-1 Experimental conditions for AOM coagulation

Sample Description	Mixing		Flocculation		Settling	Filtration
	G [s^{-1}]	t [s]	G [s^{-1}]	t [min]	[min]	
Settled	1100	20	45	15	20	-
Filtered	1100	20	45	15	20	0.45 μm

Experiments were performed for a range of coagulant concentrations from 0.5 to 20 mg Fe(III)/L at pH 5 and 8. Both settled and filtered samples were analyzed to separate the effect of coagulation from filtration on produced water quality.

7.2.4.2 Sample preparation

The suite of analytical parameters that was investigated for settled and filtered samples, along with the required volume for each parameter are presented in Table 7-2.

Table 7-2 Analytical parameters to evaluate coagulation efficiency and required sample volume

Parameter	Settled	Filtered	Volume [mL]
DOC	-	X	40
UV_{254}	-	X	100
F-EEM	X	X	40
Biopolymers	-	X	40
$TEP_{0.4}$	X	X	1000
Turbidity	X	X	100
Fe	X	X	100
$MFI\text{-}UF_{10kDa}$	X	X	60
		Total	~ 1500

The maximum volume that can be withdrawn from each jar without disturbing the settled floc blanket is 500 mL. Total volume required for all analyses per experimental condition of dose and pH is approximately 1.5 L. Therefore for each experimental condition, 3 jars (i.e., 3 L of feed solution) were coagulated. Sample preparation is schematically presented in Figure 7-1.

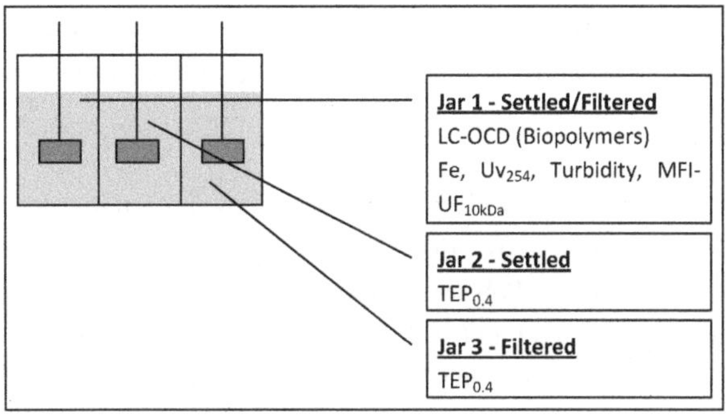

Figure 7-1 Sample preparation for analysis of settled and filtered water

7.2.5 Conventional surrogate parameters for organic matter

7.2.5.1 DOC/TOC

TOC analysis was carried out with a Shimadzu TOC analyzer. The non-purgeable organic carbon (NPOC), fraction of carbon that is not removed by gas stripping, was measured in the laboratory by acidification and purging in every step. However, the fraction of purgeable organic carbon is usually negligible. Thus for convenience NPOC is taken as TOC. Acidification is used to minimize the influence of inorganic carbon like HCO_3^-.

7.2.5.2 UV_{254}

UV_{254} absorbance was measured using a UV-2501 PC spectrophotometer (SHIMADZU, USA). QS 1 cm quartz cells were used for this wavelength. Prewashed 0.45 μm membrane filters were used to remove particulate matter prior to UV_{254} analysis. Therefore, UV_{254} measurements were only performed for filtered samples.

7.2.5.3 SUVA

SUVA was calculated from DOC and UV_{254} data obtained from the procedures described above. UV_{254} absorbance of the sample in cm^{-1} is divided by the DOC of the sample and reported in units of L/mg.m,

$$SUVA\ [L/mg.m] = UV_{254}\ [cm^{-1}] / DOC\ [mg/L] * 100\ cm/m$$

7.2.6 Advanced surrogate parameters

7.2.6.1 Fluorescence excitation-emission matrix (F-EEM)

F-EEMs were obtained using FuloroMax-3 Spectrofluorometer (HORIBA Jobin Yvon, Inc., USA), and a 4 mL, 1 cm path length cuvette. Emission spectra were generated for each sample by scanning over excitation wavelengths between 240 and 450 nm at intervals of 10 nm and emission wavelengths between 290 and 500 nm at intervals of 2 nm. The bandwidths on excitation and emission modes were both set at 1 nm. An EEM of the Milli-Q (blank) was obtained and subtracted from the EEM of each sample in order to remove most of the Raman scatter peaks. Each blank-subtracted EEM was multiplied by the respective dilution factor if any and Raman-normalized by dividing by the integrated area under the Raman scatter peak (excitation wavelength of 350nm) of the corresponding Milli-Q water and the fluorescence intensities were reported.

Protein-like organic matter exhibits a dominant peak at lower excitation/emission wavelengths while humic/fulvic substances show dominant primary and secondary peaks at higher excitation/emission wavelengths. Polysaccharides do not fluoresce when excited by light; therefore they are best represented by LC-OCD analysis. Table 7-3 shows according to fragments from pyrolysis studies, typical fluorescence responses for various compounds.

Table 7-3 Typical excitation-emission matrix peak values (Leenheer and Croué, 2003)

Description	Fluorescence range [nm]	
	Excitation	Emission
Humic-like primary peak (humic peak)	330-350	420-480
Humic-like secondary peak (fulvic peak)	250-260	380-480
Marine humic-like peak	310-330	380-420
Amino acid-like, tyrosine peak	270-280	300-320
Amino acid-like, tryptophan peak	270-280	320-350
Protein -like (albumin) peak	280	320

7.2.6.2 Liquid chromatography - organic carbon detection (LC-OCD)

Size-exclusion chromatography in combination with organic carbon detection (SEC-OCD) is an established method to separate the pool of NOM into major fractions of different sizes and chemical functions and to quantify these on the basis of organic carbon. In the LC-OCD approach, the size exclusion column (SEC) is followed by multi-detection of organic carbon (OCD), UV absorbance at 254 nm (UVD) and organic bound nitrogen (OND). Chromatograms are processed on the basis of area integration using the programme ChromCALC.

The method can give both quantitative and qualitative information of NOM with up to 10 classes of natural organics. LC-OCD separates NOM according to size/molecular weight (MW) ranging from high to low MW. The definitions, size range and retention times formally assigned to different fractions of organic carbon by DOC-Labor are presented in Table 7-4.

Biopolymers concentration is a suitable surrogate for quantifying coagulation efficiency of AOM. It represents a fraction of organic matter which is mainly composed of polysaccharides and proteins. Algal organic matter comprises, for a large part, acid polysaccharides and glycoproteins. The standard LC-OCD protocol includes sample filtration through 0.45 μm filter pore size. Hence, only filtered samples were analyzed by LC-OCD. Analyses were performed at DOC-Labor (Karlsruhe, Germany). The limit of detection of the method is compound specific and ranges from 1 - 50 μg C/L.

Table 7-4 Organic carbon fractions and descriptions of the LC-OCD method

Fraction		Size [Da]	Description
DOC: Dissolved organic carbon			Determined in the column bypass after inline 0.45μm filtration.
HOC: Hydrophobic organic carbon			DOC minus CDOC, all OC retained on the column is defined as hydrophobic. This could be natural hydrocarbons or sparingly soluble "humins".
CDOC: Chromatographable organic carbon			Obtained by area integration of the total chromatogram.
Refractory OM	HS: Humic substances	~ 1000	Based on retention time, peak shape and UV absorbance.
	BB: Building Blocks	300-500	Sub-units of HS and considered to be natural breakdown products of humics.
Biogenic OM	BP: Biopolymers	> 20,000	Very high in MW, hydrophilic, not UV-absorbing; typically polysaccharides but may also contain proteinic matter (quantified on basis of OND), and aminosugars
	LMWN: Low molecular weight neutrals	< 350	LMW weakly or uncharged hydrophilic or slightly hydrophobic (amphiphilic) compounds appear in this fraction, e.g., alcohols, aldehydes, ketones, and amino acids.
	LMWA: Low molecular weight acids	< 350	Aliphatic, LMW organic acids co-elute due to an ion chromatographic effect. A small amount of HS may fall into this fraction and is subtracted on the basis of SUVA ratios.

7.2.6.3 Alcian blue staining and $TEP_{0.4}$ measurement

The method proposed by Passow and Alldredge (1995) and modified by Villacorte et al. (2013) was used to measure $TEP_{0.4}$. Samples were filtered under low, constant vacuum onto 0.4μm polycarbonate filters (Whatman Nuclepore) and stained with Alcian blue. Alcian blue is a dye that binds specifically with anionic acidic polysaccharides and glycoproteins (Ramus, 1977). The dye bound to particles is re-dissolved in sulphuric acid and the amount eluted is quantified colorimetrically. Blank measurements were performed to correct for dye taken up by filters. The absorbance was measured with UV-vis spectrophotometer in a 1 cm cuvette against Milli-Q water as reference. The absorbance maximum of the solution lies at 787 nm.

To eliminate the impact of salinity on Alcian blue effectiveness, 1 mL of Milli-Q water was filtered through the TEP gel layer prior to staining, to rinse the saline water. Calibration was performed with Xanthan gum and TEP concentrations are expressed as mg Xanthan equivalent per litre (mg Xeq/L). Limit of detection of the method is 0.015 mg Xeq/L.

It must be noted that significant amounts of acid polysaccharides in surface waters are colloidal TEP, which are smaller than 0.4 μm (Villacorte, 2009). Hence measurements of TEP_{10kDa} can provide valuable insight on the removal of smaller AOM/TEP. At the time this study was conducted the TEP_{10kDa} method was still under development and not fully validated. TEP concentrations harvested from different algal cultures turned out to be affected by storage (Villacorte, 2014). Storage of samples was inevitable and in most cases in the same order of length, which means that the results presented are not absolute in level, but relative.

7.2.7 Surrogate parameters non-directly related to AOM

7.2.7.1 Iron

Iron analysis was done using the colorimetric phenanthroline method (Sandell, 1959). Measurement of iron using advanced techniques such as ICP-MS or AAS was not feasible due to the high salinity of the water samples that can damage the instruments. However, the phenanthroline method proved to be reliable for measuring in high salinity, providing good calibration curves and a low detection limit of 10 μg/L.

The procedure requires boiling the samples with concentrated HCl and NH_2OH, to reduce all iron to Fe^{2+}, which reacts with phenanthroline and is measured at a wavelength of 510 nm. All glassware used for the analyses were cleaned with concentrated HCl and rinsed with Milli-Q water prior to each measurement. Samples were measured immediately after collection, or if storage was necessary, samples were acidified with 2% HCl and kept at 4 °C.

7.2.7.2 Turbidity

Turbidity analysis was carried out with a HACH-Lange turbidity meter (Germany). The typical 90° angle was used to determine turbidity (NTU) at a wavelength of 860 nm. The samples were measured immediately after collection at room temperature. Limit of detection of the method is 0.05 NTU.

7.2.7.3 Modified Fouling Index - Ultrafiltration (MFI-UF)

The Modified Fouling Index (MFI) is an index for quantifying the fouling potential - due to particulate matter - of RO feed water (Schippers and Verdouw, 1980). The principle of this index is based on the occurrence of a cake/gel layer during filtration at constant pressure, through membranes having pores of 0.45 or 0.05 μm. An improvement to the method, the Modified Fouling Index - Ultrafiltration (MFI-UF) at constant filtration flux, and making use of membrane filters with much smaller pores (Boerlage et al., 2004; Salinas Rodriguez, 2011) was used in this study. Flat sheet, regenerated cellulose (RC) membranes (Millipore, USA) of 25 mm diameter and a MWCO of 10 kDa were used for the experiments. The membranes were soaked in Milli-Q water to wet the pores and remove the preserving coat. Filtration flux for $MFI\text{-}UF_{10kDa}$ should resemble flux values in seawater RO membranes, i.e., 15 L/m^2h. This low flux could not be applied in the experimental setup for MFI-UF measurements, as the signal to noise ratio is very low at low flux values. Hence, $MFI\text{-}UF_{10kDa}$ was performed at a filtration flux of 60 L/m^2h. Filtration volume was approximately 60 mL. Limit of detection of the method is 50 s/L^2.

7.3 Results and discussion

7.3.1 AOM characterization

Harvested AOM was analyzed by LC-OCD to verify the concentration of various fractions of organic matter. According to calibration with various standards, elution time is around 30 min, 43.4 min, 50 min, 52 min and 55 min, for biopolymers, humic substances, building blocks, LMW acids and LMW neutrals respectively (Figure 7-2).

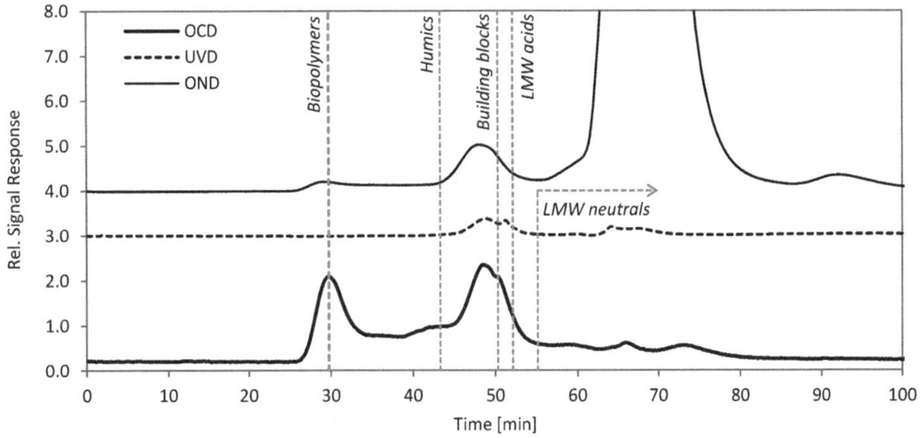

Figure 7-2 LC-OCD chromatogram (with inline 0.45 μm filtration) of AOM stock solution (Batch 1)

The first peak corresponds to the biopolymers fraction (elution time 30 min). The upper size limit of this fraction is defined by the exclusion limit of the SEC column, while the lower size limit is defined by the left slope of the HS fraction. The rapid elution of this fraction reflects high molecular weight (>> 10 kDa, column separation range), hydrophilic material. The fraction does not absorb UVD_{254}, and is therefore lacking aromatic and unsaturated structures. The presence of nitrogen, as reflected by the OND signal, indicates that the material contains proteins or amino sugars. These observations are characteristic of the presence of polysaccharides and proteins (Huber et al., 2011). TEP, which is excreted copiously by *Chaetoceros affinis*, is mostly comprised of acidic polysaccharides and glycoproteins.

The presence of low molecular weight acids may be attributed to the medium in which the diatoms were grown or debris and breakdown products of algal cell walls. UVD signal detected for these substances further confirms their aromatic, unsaturated nature. OND peak observed at elution time > 60 min indicates the presence of inorganic species, e.g., nitrate and ammonium (Huber et al., 2011).

Preparing AOM stock solutions in different batches - which was inevitable - may be disadvantageous in terms of concentration and characteristics of different fractions of organic matter. LC-OCD of AOM stock solutions obtained from two distinct batches revealed small differences in concentration of DOC and biopolymers; 4% and 8% respectively. AOM batch 2 was only used for $TEP_{0.4}$ measurement at pH 5. All other experiments were performed with the same batch of AOM, i.e., batch 1.

7.3.2 Feed water characterization

Feed solutions were prepared by diluting AOM stock solutions in synthetic seawater. Samples of feed water and blank synthetic seawater were characterized with measurements of DOC, LC-OCD, and UV_{254} (Figure 7-3). The aim of these measurements was to identify the relative suitability of different surrogate parameters in characterizing AOM solutions in seawater.

DOC concentration of the feed water is approximately 1.9 mg C/L as measured by LC-OCD. Blank measurement of synthetic seawater reflects approximately 0.6 mg C/L. The DOC content of synthetic seawater is very close to the limit of detection of the instrument, i.e., 0.5 mg C/L.

Characterizing removal efficiency of AOM based on DOC concentrations is of limited value, mainly for two reasons. First, DOC is a bulk parameter that does not reflect variations in the different fractions of organic matter. An important part does not belong to the category of polymers in which we are interested. Second, to report target removal rates of > 90%, feed DOC concentration should be at least 6 mg C/L. This has to do with the limit of detection (LoD) of the measurement which is approximately 0.5 mg C/L. The high LoD in these measurements is attributed to the high salinity (~ 35%) of the background solution which makes detection of organic carbon more tedious.

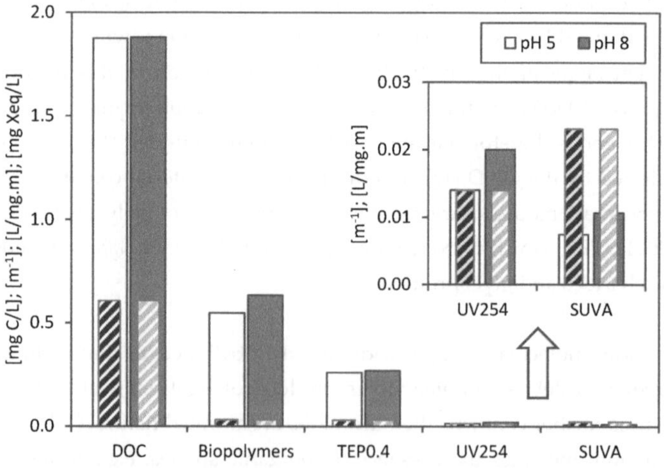

Figure 7-3 Characterization in terms of DOC-related parameters of feed water (solid bars), and blank synthetic seawater (striped bars)

Note that the effect of coagulation on organic matter in fresh water primarily depends on pH, coagulant dose and concentration of organic matter. The lower the pH and the higher the dose, the better the removal. This is attributed to the dominant presence of humic substances. In this study, biopolymer concentration was fixed to the highest value encountered during real bloom conditions in North Sea water (Oosterschelde, The Netherlands) in 2009.

Biopolymers concentration was measured at 0.55 and 0.63 mg C/L for filtered feed solutions at pH 5 and 8 respectively. However, it is not unlikely that the difference in concentration stems from the standard deviation of the LC-OCD measurement. This feed concentration falls well within the range of biopolymers observed in seawater during algal bloom conditions. Biopolymers concentration of 0.5 mg C/L was measured in North Seawater during a *Phaeocystis* bloom in 2009 (Villacorte et al., 2010).

Biopolymer concentration in synthetic seawater (blank) was approximately 0.03 mg C/L. The low concentration of blank means that for feed water concentrations as low as 0.5 mg C/L, > 90% removal rate can be reported. LC-OCD analysis of Milli-Q water prior to salt addition did not reflect peaks for humic substances, building blocks, LMW acids and LMW neutrals.

UV_{254} values of 0.014 and 0.02 m^{-1} were measured for filtered feed water samples at pH 5 and pH 8 respectively. This falls at the level of blank measurements for synthetic seawater (~ 0.015 m^{-1}).

SUVA Corresponding SUVA values are 0.007 and 0.01 L/mg.m. For comparison, plants treating water with SUVA < 2 are exempt from enhanced coagulation (USEPA, 1998). SUVA values for blank synthetic seawater are higher than feed samples as the DOC content is lower. UV absorbance at 254 nm is used as a surrogate for DOC, reflecting aromatic and saturated structures (Chang et al., 2000).

For AOM, which is mostly composed of polysaccharides and glycoproteins, measurement of UV_{254} and/or SUVA is not a suitable surrogate. However, certain UV absorbing compounds (e.g., iron) are found to interfere with UV measurements (Dilling and Kaiser, 2002; Weishaar et al., 2003) particularly for low DOC waters. Hence, in this study, UV_{254} measurements can provide information on the presence of colloidal iron in treated water samples.

$TEP_{0.4}$ of the feed water was approximately 0.27 mg Xeq/L. The limit of detection of the method is 0.015 mg Xeq/L, implying that $TEP_{0.4}$ removal of at least 90% can be reported. Synthetic seawater blanks showed levels close to limit of detection, approximately 0.03 mg Xeq/L.

F-EEM of blanks (synthetic seawater) and feed water (with and without filtration) are presented below. pH did not seem to affect the peaks and intensities of the samples. Filtration reduced the protein-like peak to levels close to blank values.

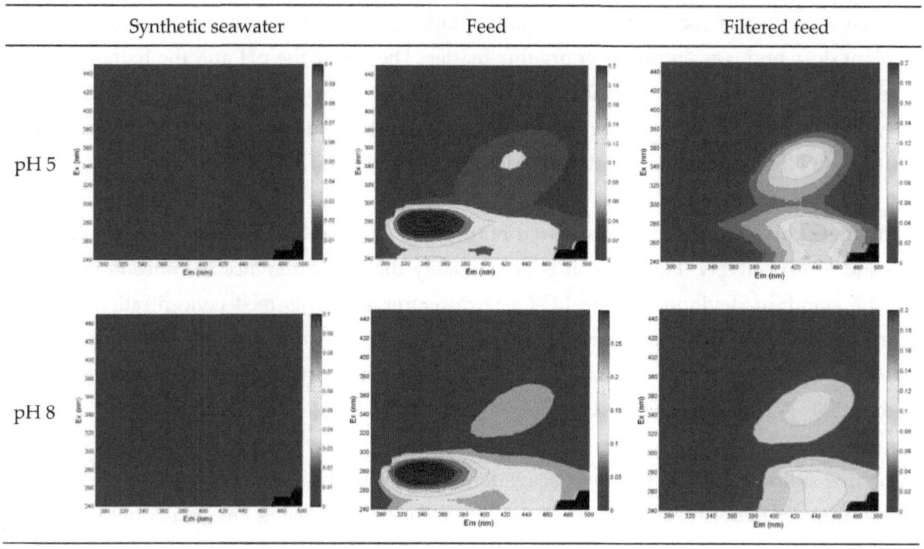

7.3.3 Effect of coagulation followed by settling and filtration

This section investigates the effect of coagulant dose and pH on the removal of AOM obtained from *Chaetoceros affinis* in solutions of synthetic seawater.

7.3.3.1 DOC, UV_{254}, SUVA

The effect of coagulation followed by settling and filtration on AOM removal in terms of conventional surrogate parameters for organic matter characterization (i.e., DOC, UV_{254} and SUVA) is presented in Figure 7-4. At pH 5, DOC concentration decreased from 1.88 mg C/L to 1.32 mg C/L for increase in coagulant dose up to 10 mg Fe(III)/L. Thereafter a slight increase in DOC was observed for coagulation at 20 mg Fe(III)/L.

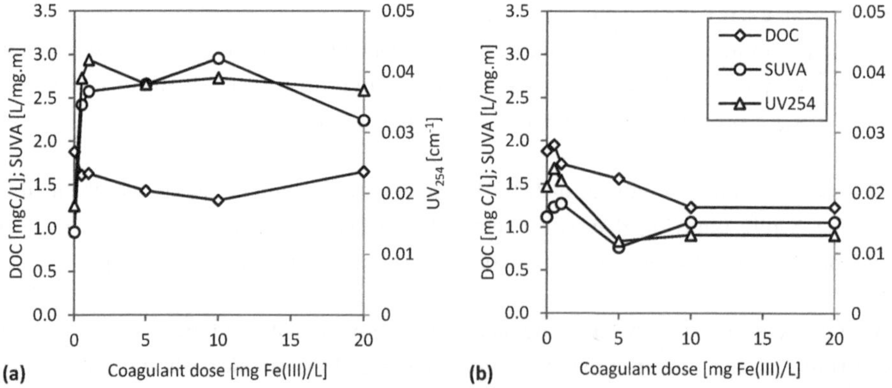

Figure 7-4 Organic matter parameters as a function of coagulant dose for filtered samples at (a) pH 5, and (b) pH 8

At pH 8, DOC concentration declined with increase in coagulant dose from 1.88 mg C/L to 1.23 mg C/L. UV_{254} and SUVA values increased with coagulant addition at pH 5. SUVA increased from 0.7 L/mg.m for no coagulant addition to 2.4 L/mg.m at 0.5 mg Fe(III)/L. Further addition of coagulant (up to 20 mg Fe(III)/L) did not have a significant impact on SUVA. Similarly for UV_{254}, addition of coagulant resulted in an increase in absorbance from 0.01 cm^{-1} for no coagulant addition to 0.04 cm^{-1} at 0.5 mg Fe(III)/L. At pH 8, UV_{254} absorbance was slightly reduced for coagulant concentrations > 5 mg Fe(III)/L. However, SUVA values remained unaffected by coagulant addition, i.e., 1.1 ± 0.2 L/mg.m for 0 - 20 mg Fe(III)/L.

7.3.3.2 F-EEM

Two types of DOM fluorescence signals were observed in the spectra as humic-like (humic acids and fulvic acids) and a protein- or amino acid-like fluorescence. The protein-like peak may be attributed to glycoproteins present in TEP. Coagulation at concentrations less than 1 mg Fe(III)/L did not affect the removal of protein-like substances (Figure 7-5). For coagulant dose ≥ 5 mg Fe(III)/L, complete removal was observed at pH 5, while removal was approximately 70% at pH 8. Filtration through 0.45 μm effectively reduced protein-like peak intensities to levels comparable with synthetic seawater blank measurements, regardless of coagulant dose and pH.

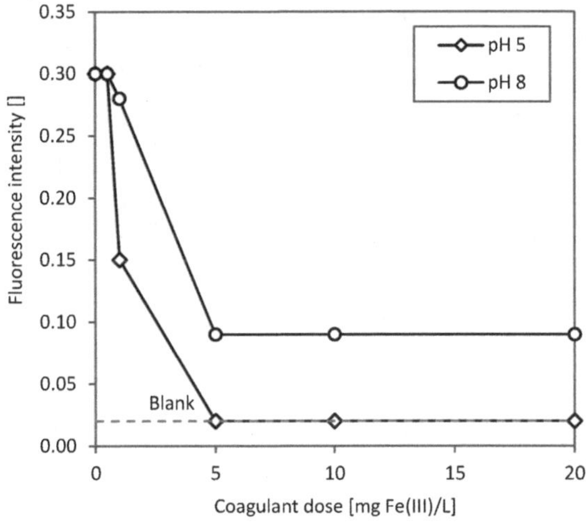

Figure 7-5 F-EEM spectra as a function of coagulant dose and pH for settled water samples

Humic-like fluorescence occurs at 420-450 nm from excitation at 230-260 and 320-350 nm (primary and secondary peaks). The protein-like fluorescence arises from the fluorescence of aromatic amino acids, either free or as protein constituents, and is observed at an emission of 300-305 nm (tyrosine-like) and 340-350 nm (tryptophan-like) from excitation at 220 and 275 nm. Polysaccharides do not fluoresce when excited by light; therefore other techniques such as LC-OCD and TEP measurements are required for their detection and quantification.

7.3.3.3 LC-OCD

LC-OCD chromatograms of AOM solutions for different coagulant dose at pH 5 and 8 are presented in Figure 7-6. Only filtered samples were analysed by LC-OCD. Coagulation resulted in a significant reduction in biopolymers peak for both pH values. Reduction was particularly significant at pH 8 for coagulant dose of 5 mg Fe(III)/L and above.

The second peak in the chromatograms, corresponding to low molecular weight acids, did not show a considerable decline in signal response as a function of coagulant dose, regardless of pH. The efficiency of removing organic carbon is proportional to molecular size, with larger molecular weight components more effectively removed than small ones (Dennett et al., 1996).

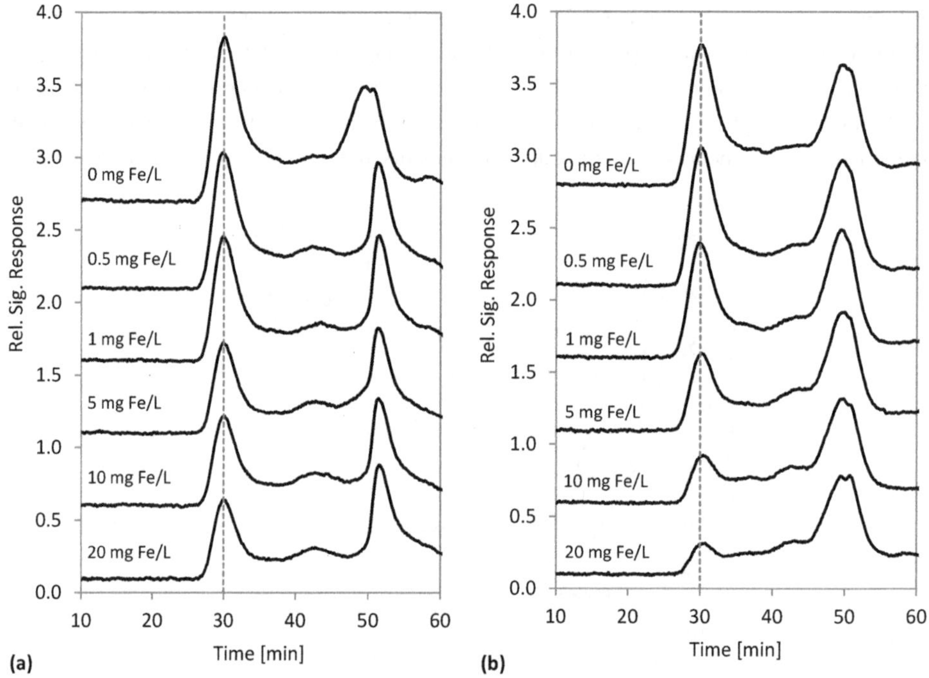

Figure 7-6 Organic carbon chromatograms for treated samples as a function of coagulant dose at (a) pH 5 and (b) pH 8; biopolymers elution time 30 min

Quantitative data are presented in Figure 7-7, showing higher removal of biopolymers at high coagulant dose. Removal rate at pH 8 was relatively higher than at pH 5. In freshwater coagulation, the opposite trend is observed, namely, higher DOC removal at low pH. This is attributed to the dominant presence of humic substances. The relatively high concentration of biopolymers in algal bloom-impaired seawater might explain the observed difference in behaviour.

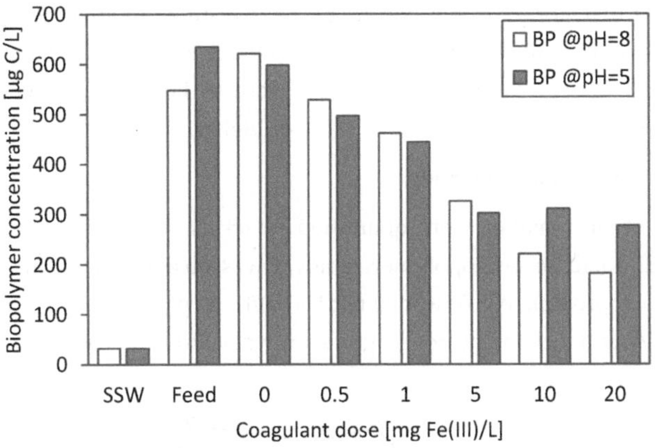

Figure 7-7 Biopolymer concentration as a function of coagulant dose at pH 5 and 8

7.3.3.4 $TEP_{0.4}$

$TEP_{0.4}$ concentrations of settled and filtered samples as a function of coagulant dose at pH 5 and 8 are shown in Figure 7-8. Increase in coagulant dose resulted in a reduction in $TEP_{0.4}$ concentrations for settled samples at both pH values. $TEP_{0.4}$ concentration in filtered samples was below detection limit (0.015 mg Xeq/L) regardless of coagulant dose and pH for 95% confidence interval. This was expected as the samples were filtered through 0.45 µm filters.

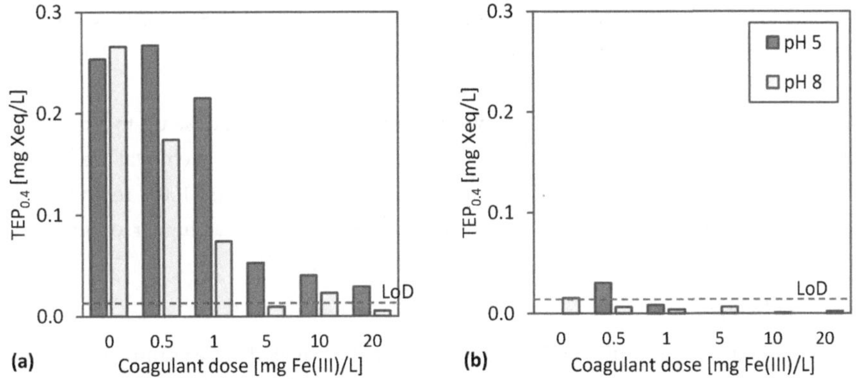

Figure 7-8 $TEP_{0.4}$ concentrations in mg Xeq/L as a function of coagulant dose and pH for (a) settled and (b) filtered samples

TEP measurements were performed by filtering the solutions through polysulphone membranes with pore size of 0.4 µm and therefore it was not expected to find a concentration of $TEP_{0.4}$ in filtered samples. Ideally filtered samples should be prepared by filtering settled solutions through a range of pore sizes for a thorough understanding.

Taking into account that storage of samples might have resulted in lower levels than real levels, the removal rates are most likely higher than those observed. For coagulated samples (≥ 5 mg Fe(III)/L, pH 8) and filtered samples, levels were close to LoD. Consequently, removal rates could be determined as (1 - LoD/0.27) x 100% = 95% or higher.

7.3.3.5 Algal organic matter removal

Removal rates of organic matter were quantified in terms of biopolymers, $TEP_{0.4}$ and DOC (Figure 7-9). Steady increase in biopolymer removal was observed with increase in coagulant dose at pH 8, up to approximately 70% at 20 mg Fe(III)/L. At pH 5, removal efficiency was about 55% at 5 mg Fe(III)/L and did not improve significantly beyond this dose.

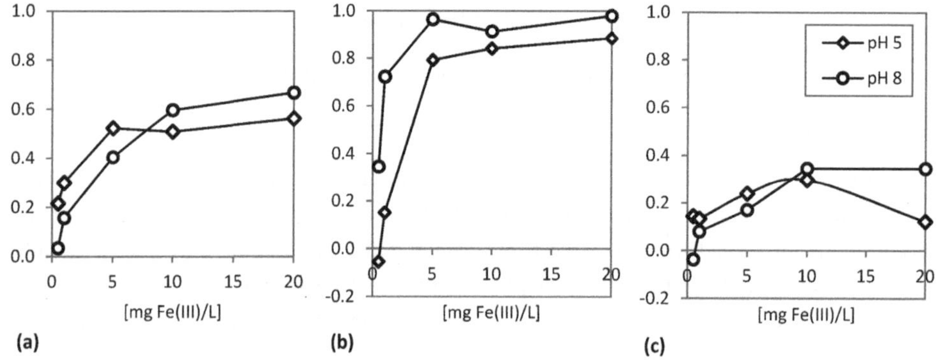

Figure 7-9 Removal rates of (a) biopolymers, (b) $TEP_{0.4}$ (settled samples) and (c) DOC as a function of coagulant dose and pH

$TEP_{0.4}$ removal was substantially improved with increase in coagulant dose up to 5 mg Fe(III)/L, beyond which further improvement in removal was marginal. $TEP_{0.4}$ removal was relatively higher at pH 8 than pH 5. Note that $TEP_{0.4}$ for filtered samples showed 100% removal for 95% confidence interval (Figure 7-8). DOC removal at pH 8 was higher at higher coagulant dose. However, at pH 5 a trend in removal of DOC could not be deduced.

7.3.3.6 $MFI\text{-}UF_{10kDa}$

Effect of coagulant dose and pH on the fouling potential of treated water samples as measured by $MFI\text{-}UF_{10kDa}$ was studied (Figure 7-10). For coagulation at pH 8, $MFI\text{-}UF_{10kDa}$ of settled water increased slightly when low coagulant dose (i.e., 0.5 mg Fe(III)/L) was applied.

Coagulation of hydrophilic organic compounds (e.g., AOM and associated TEP) at low coagulant dose is expected to be governed by reactions between positively charged iron species and the negative functional groups on the biopolymers. This might result in compounds having a lower or a higher specific fouling potential (MFI-UF). With increase in coagulant dose, enmeshment is most likely the predominant mechanism, and $MFI\text{-}UF_{10kDa}$ values decrease significantly; by 45% and 90% for coagulant dose of 1 and 20 mg Fe(III)/L respectively.

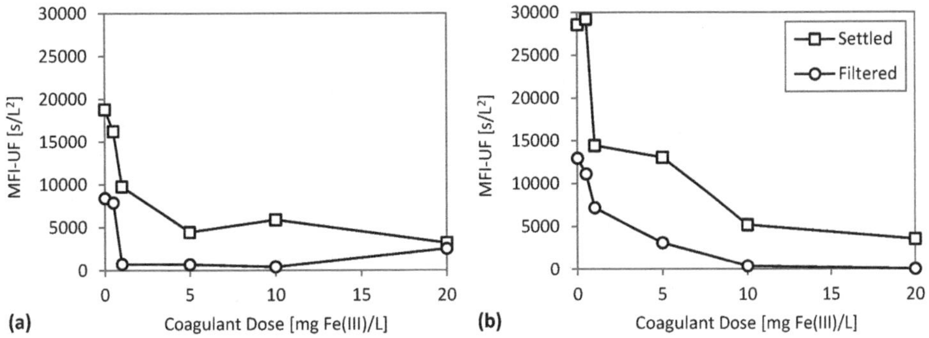

Figure 7-10 MFI-UF$_{10kDa}$ of treated water as a function of coagulant dose at (a) pH 5 and (b) pH 8

Filtration through 0.45 μm resulted in a 50% decrease in MFI-UF$_{10kDa}$. Addition of coagulant resulted in a steady decline in MFI-UF$_{10kDa}$, from 14,000 s/L^2 for no coagulant addition to below detection limit for ≥ 10 mg Fe(III)/L. At pH 5, MFI-UF$_{10kDa}$ declined steadily with increase in coagulant dose up to 5 mg Fe(III)/L for settled water, after which MFI-UF$_{10kDa}$ was stable at around 4000 s/L^2. MFI-UF$_{10kDa}$ of filtered water was strongly reduced for coagulation at > 1 mg Fe(III)/L. A slight increase in MFI-UF$_{10kDa}$ was observed for the higher coagulant dose of 20 mg Fe(III)/L.

MFI-UF$_{10kDa}$ values at pH 5 were found to be consistently lower than those at pH 8. This included MFI-UF$_{10kDa}$ of feed water, when no coagulant was added. However, normalized MFI-UF$_{10kDa}$ values at pH 8 and 5, revealed comparable removal rates and declining trends in both settled and filtered samples. Lower MFI-UF values observed at lower pH, could be attributed to three distinct aspects. First, pH may affect the structure of AOM and consequently the specific fouling potential. Effect of pH on MFI-UF values of AOM solutions at 0.5 mg C/L as biopolymers was investigated (Figure 7-11). MFI-UF values decreased by more than 30% from pH 8 to pH 5.

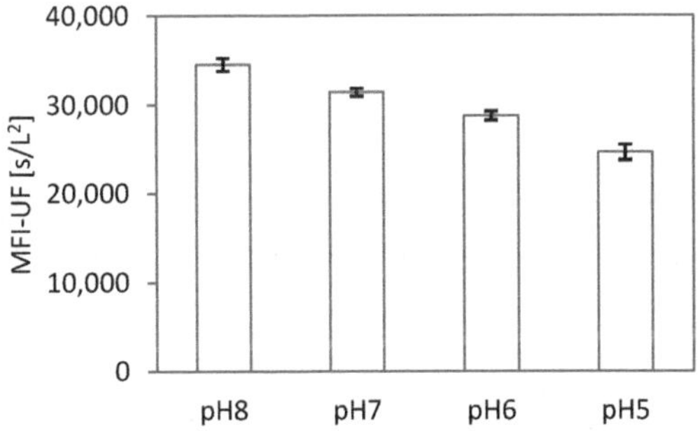

Figure 7-11 Effect of pH on MFI-UF$_{10kDa}$ of AOM at 0.5 mg C/L as biopolymers

Second, pH can affect the membrane resistance by shrinking or enlarging pores or surface characteristics that will result in higher retention of the foulants. Effect of pH on membrane resistance was investigated by filtering SSW at pH 5 and 8 through 10 kDa RC membranes. Membrane resistance was measured at 4.5 $*10^{12}$ and 4.6 $*10^{12}$ for pH 5 and 8 respectively. Resistance at pH 5 seems to be slightly lower than that at pH 8. However, each measurement was only performed once making it difficult to conclude whether a statistical difference exists between the two pH values.

Third, membrane resistance of 10 kDa RC filters used in MFI-UF measurements differs per membrane batch and within batches. Experiments at pH 5 and pH 8 were performed with 2 distinct batches of RC filters (10 filters in each box). Membrane resistance (R_m) were calculated based on data obtained from Milli-Q filtration (according to procedure for cleaning filters prior to sample filtration) and are presented in Figure 7-12.

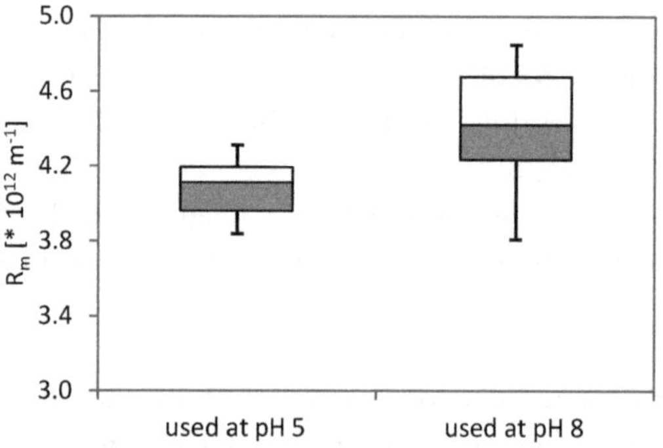

Figure 7-12 Membrane resistance (R_m) of RC filters used at different pH values

Experiments at each pH value were performed with a distinct batch of filters. Average R_m (of 14 data points) used at pH 5 was 4.1 $*10^{12} \pm 3.8\%$ and that at pH 8 was 4.4 $*10^{12} \pm 6.7\%$. This difference is significant and might explain the lower MFI-UF values at pH 5. This observation indicates that MFI-UF at pH 5 could be lower due to enlarging of the pores.

7.3.3.7 Iron

Iron fouling might cause a performance loss of the membrane system, in terms of loss of permeability. RO membrane manufacturers recommend RO/NF feed water concentrations of iron below 0.05 mg Fe(III)/L. However, they express this as total iron and do not specify the species. Residual iron concentrations in settled and filtered samples were measured in duplicate (Figure 7-13).

Iron residual in settled and filtered samples was relatively high at pH 5, exceeding the recommended limit. However, iron solubility at pH 5 is very low, i.e., 1.8 μg Fe(III)/L. Therefore, solubility of ferric hydroxide cannot explain iron residual in the samples. Slower coagulation/flocculation kinetics at lower pH may result in the formation of micro-colloids, present in large numbers, which settle slowly and can pass through 0.45 μm filters and contribute to residual iron in filtered water samples. Residual iron in settled water at pH 8 ranged from 0.5 to 1 mg Fe(III)/L. Filtration through 0.45 μm filters reduced residual iron at pH 8 to values below detection limit, i.e., 20 μg/L.

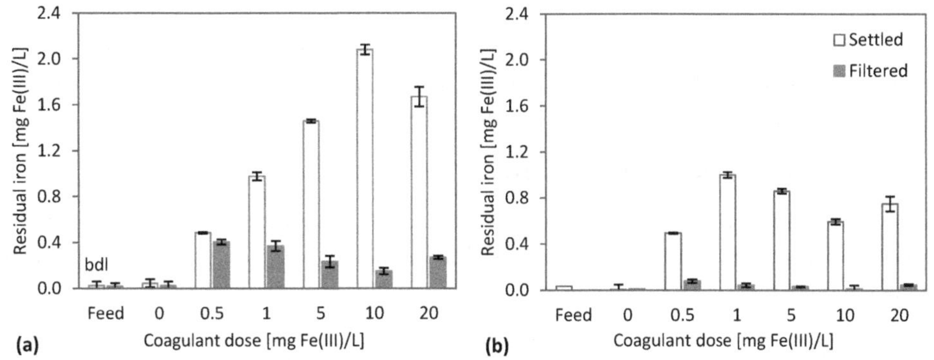

Figure 7-13 Iron residual in treated samples as a function of coagulant dose at (a) pH 5 and (b) pH 8

7.3.3.8 Turbidity

Maximum turbidity of 1.0 NTU is recommended by membrane manufacturers (Hydranautics, DOW) for SWRO feed water. As AOM are largely composed of transparent micro-particles (i.e., TEP), they are not expected to impart turbidity to the water. However AOM stock solutions were slightly turbid due to algal residues as reflected in feed water turbidity values. In treated water samples, turbidity can be due to the presence of colloidal iron and/or phytoplankton residues. Residual turbidity for coagulation at different dose and pH values is presented in Figure 7-14 for settled and filtered samples.

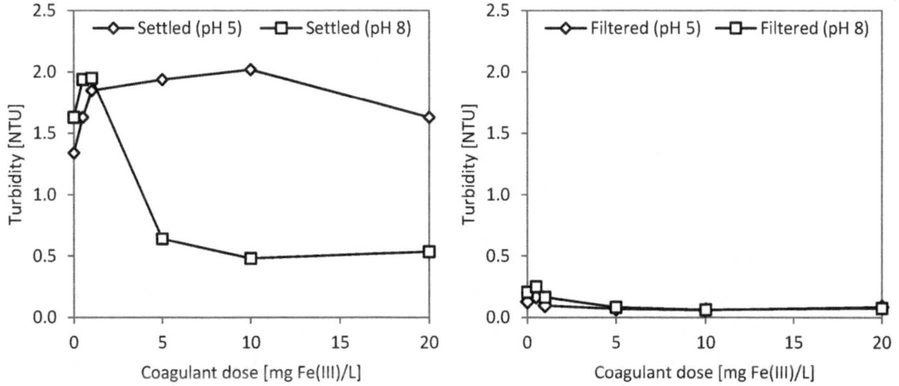

Figure 7-14 Turbidity for settled and filtered samples as a function of coagulant dose at pH 5 and 8

It is remarkable that turbidity after settling at pH 8 remained high for coagulant concentrations of 0.5 and 1 mg Fe(III)/L. The high turbidity at low coagulant concentrations may be attributed to the presence of iron micro-colloids. For coagulant dose of 5 mg Fe(III)/L and higher, visibly larger flocs (for higher coagulant dose, large flocs were clearly observed during the experiments) were formed and settling efficiency was enhanced, reflected by a lower turbidity in the settled supernatant.

At pH 5, turbidity remained high at all coagulant doses, most likely due to the presence of iron in the settled samples. Filtration through 0.45 μm membrane filters reduced turbidity in all cases - with and without coagulant dosage - to less than 1.0 NTU.

7.3.4 Coagulation mechanisms

At low coagulant dose the residual iron in treated samples was rather high. In the presence of a sufficient biopolymer concentration, iron was found in filtered samples, particularly at low pH values (Figure 7-15).

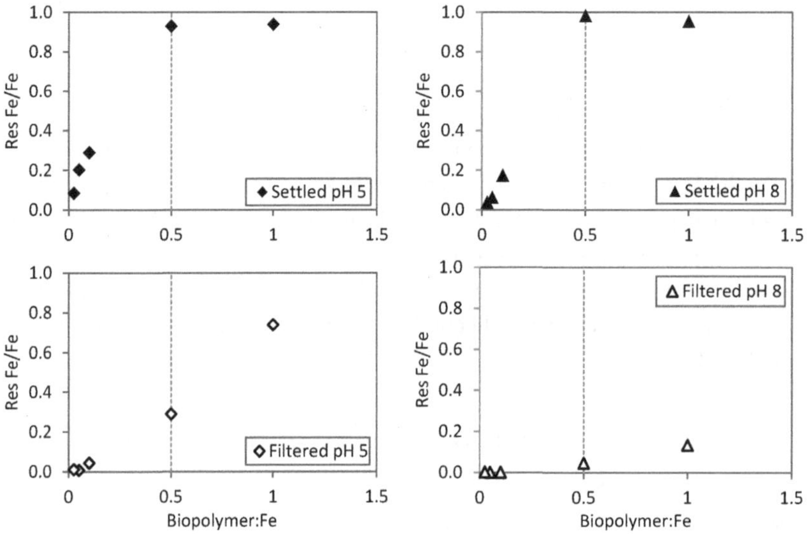

Figure 7-15 Iron residual in treated water samples as a function of biopolymer:Fe ratio and pH; Res Fe/Fe is residual iron measured in the samples over coagulant dose

This effect was more pronounced at pH 5 than at pH 8. Bernhardt et al. (1985) gave a possible explanation for these observations by suggesting - in fresh water containing algae - that when iron is added to a feed water containing polyanions of AOM, a competition reaction occurs not only amongst the polymerized iron hydroxo polymers themselves but also with the polymeric AOM molecules. AOM occupies reactive sites on iron hydroxo polymers, which at low iron concentrations leads to preventing cross linking and clustering of iron hydroxo polymers resulting in aggregation and stabilization of iron micro-colloids. These micro-colloids have a low settling velocity and can partly pass through a 0.45 μm filter and are consequently measured as 'dissolved iron'. On the other hand an excess of iron makes cross linking and clustering of the iron micro-colloids to larger aggregates possible, e.g., by enmeshment, with the result that all Fe compounds are retained on a 0.45 μm filter. At pH 5 a larger amount of micro-colloids are present in filtered samples due to lower aggregation rates and floc growth kinetics and therefore enmeshment is not very effective.

The relation between MFI-UF values and organic matter related parameters DOC, $TEP_{0.4}$ and biopolymers is shown in Figure 7-16. Statistical correlations could not be established between MFI-UF values and biopolymers and $TEP_{0.4}$ concentrations, as indicated by low R^2 values, i.e., 0.74 and 0.58 respectively. The very low R^2 with DOC might be explained by the fact that below a certain DOC level all biopolymers are removed.

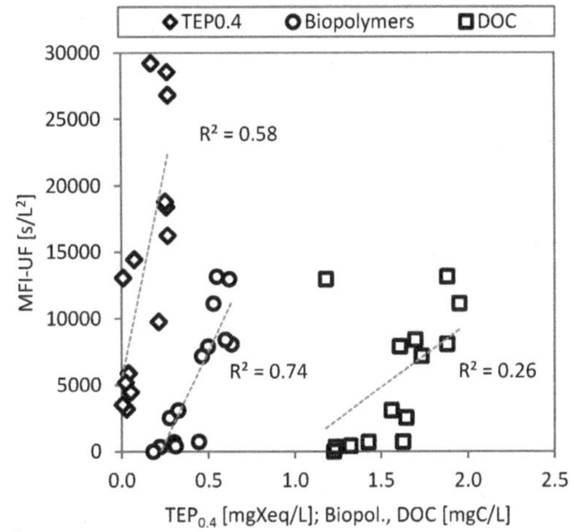

Figure 7-16 Correlation $MFI\text{-}UF_{10kDa}$ with AOM surrogates $TEP_{0.4}$, biopolymers and DOC

7.4 Conclusions

AOM harvested from laboratory cultivated marine diatoms *Chaetoceros affinis* was characterized with conventional and advanced parameters showing that about 25 to 30% of dissolved organic matter consists of biopolymers. AOM consisted mainly of compounds carrying no aromatic structure. $TEP_{0.4}$ was present in concentrations equivalent to at least 0.5 mg Xeq/L.

Coagulation followed by sedimentation was effective in reducing biopolymer concentration (as measured with LC-OCD) and fouling potential (as measured with $MFI\text{-}UF_{10kDa}$) by up to 70-80%. $TEP_{0.4}$ removal of more than 80% was achieved at coagulant dose of 5 mg Fe(III)/L or higher.

Filtration of settled samples through membranes with pores of 0.45 μm further removed $TEP_{0.4}$ (as expected) and $MFI\text{-}UF_{10kDa}$.

The effect of pH on removal rates of surrogate parameters investigated in this study was not very pronounced.

Observed residual iron concentrations and turbidity suggests that at low coagulant dose, AOM particles react with the iron hydroxo-polymers. At increasing dose enmeshment becomes more dominant.

The large suite of analytical parameters applied in this study revealed that some surrogates commonly used in quantifying coagulation efficiency of organic matter in fresh water, e.g. UV_{254} and SUVA are not suited for the characterization of coagulation efficiency of AOM in high salinity water. Advanced surrogate parameters such as $TEP_{0.4}$, biopolymers concentration as measured by LC-OCD and MFI-UF may be more accurate in characterizing the presence of AOM in feed solutions, and are therefore more suitable for studying the coagulation efficiency of these systems. Whether these advanced surrogate parameters are suitable for predicting the rate of fouling in MF/UF and RO systems remains to be investigated.

7.5 Acknowledgements

Thanks are due to Kaisa Karisalmi, Andreja Peternelj, Tarja Turki, Mehrdad Hessampour from KEMIRA; DOC-Labor (Karlsruhe); CCAP (Oban); Athambawa Meeralabbe, Yuli Ekowati, Helga Calix and Loreen Villacorte.

7.6 References

Alldredge, A.L., Passow, U., Logan, B.E., 1993. The abundance and significance of a class of large, transparent organic particles in the ocean. Deep-Sea Research I 40, 1131-1140.

Berman, T., 2010. TEP - a major challenge for water filtration. Filtration & Separation 47(2), 20-22.

Berman, T., Holenberg, M., 2005. Don't fall foul of biofilm through high TEP levels. Filtration & Separation 42(4), 30–32.

Bernhardt, H., and Hoyer, O., 1979. Characterization of organic constituents by the kinetics of chlorine consumption. In: Oxidation Techniques in Drinking Water Treatment. EPA-Report No.570/9-020, CCMS-111, 110-137.

Bernhardt, H., Hoyer, O., Schell, H., Lusse, B., 1985. Reaction mechanisms involved in the influence of algogenic matter on flocculation. Zeitschrift fur Wasser und Abwasser forschung 18, 18-30.

Boerlage, S.F.E., Kennedy, M., Tarawneh, Z., de Faber, R., Schippers, J.C., 2004. Development of the MFI-UF in constant flux filtration. Desalination 161, 103-113.

Dennett, K.E., Amirtharajah, A., Moran, T.F., Gould, J.P., 1996. Coagulation: its effect on organic matter. Journal of AWWA 84(4), 129-142.

Dilling, J., Kaiser, K., 2002. Estimation of the hydrophobic fraction of dissolved organic matter in water samples using UV photometry. Water Research 36, 5037-5044.

Fogg, G.E., 1983. The ecological significance of extracellular products of phytoplankton photosynthesis. Bot. Mar. 26, 3–14.

Guillard, R.R., Ryther, J.H., 1962. Studies on marine planktonic diatoms. I. Cyclotella nana Hustedt and Detonula confervacaea (Cleve) Gran. Canadian Journal of Microbiology 8, 229-239.

Henderson, R.K., Baker, A., Parsons, S.A., Jefferson, B., 2008. Characterisation of algogenic organic matter extracted from cyanobacteria, green algae and diatoms. Water Research 42, 3435-3445.

Henderson, R.K., Parsons, S.A., Jefferson, B., 2010. The impact of differing cell and algogenic organic matter (AOM) characteristics on the coagulation and flotation of algae. Water Research 44, 3617-3624

Her, N., Amy, G., Park, H.-R., Song, M., 2004. Characterizing algogenic organic matter (AOM) and evaluating associated NF membrane fouling. Water Research 38(6), 1427-1438.

Huber, S. A., Balz, A., Abert, M. and Pronk, W., 2011. Characterisation of aquatic humic and non-humic matter with size-exclusion chromatography – organic carbon detection – organic nitrogen detection (LC-OCD-OND). Water Research 45(2), 879-885.

Joret, J.C., Levi, Y., Dupin, T., Gibert, M., 1988. Rapid method for estimating bioeliminable organic carbon in water. In: Proceedings of the AWWA Annual Conference and Exposition. Denver, Colorado, 1715-1725.

Mopper, K., Zhou, J., Sri Ramana, K., Passow, U., Dam, H.G., Drapeau, D.T., 1995. The role of surface-active carbohydrates in the flocculation of a diatom bloom in a mesocosm. Deep-Sea Research II 42(1), 47-73.

Myklestad, S.M., 1995. Release of extracellular products by phytoplankton with special emphasis on polysaccharides. Science of the Total Environment 165, 155-164.

Passow, U., & Alldredge, A.L., 1995. A dye-binding assay for the spectrophotometric measurement of transparent exopolymer particles (TEP). Limnology and Oceanography 40, 1326–1335.

Passow, U., 2002. Transparent exopolymer particles (TEP) in aquatic environments. Progress in Oceanography 55, 287-333.

Ramus, J., 1977. Alcian blue: a quantitative aqueous assay for algal acid and sulphated polysaccharides. Journal of Phycology 13, 345-348.

Salinas Rodriguez, S.G., 2011. Particulate and organic matter fouling of SWRO systems: characterization, modelling and applications. PhD thesis, UNESCO-IHE/TUDelft, Delft.

Sandell, E.B., Colorimetric Determination of Traces of Metals, 3rd Ed. Interscience Publishers, Inc., New York, 1959.

Schippers, J.C., and Verdouw, J., 1980. The modified fouling index, a method of determining the fouling characteristics of water. Desalination 32, 137-148.

van der Kooij, D., Visser, A., Hijnen, W.A.M., 1982. Determination of easily assimilable organic carbon in drinking water. Journal of the American Water Works Association 74, 540-545.

Villacorte, L.O., Ekowati, Y., Winters, H., Amy, G.L., Schippers, J.C., Kennedy, M.D., 2013. Characterisation of transparent exopolymer particles (TEP) produced during algal bloom: a membrane treatment perspective. Desalination and Water Treatment 51 (4-6), 1021-1033.

Villacorte, L.O., Schurer, R., Kennedy, M., Amy, G., Schippers, J.C., 2010. The fate of transparent exopolymer particles in integrated membrane systems: a pilot plant study in Zeeland, The Netherlands. Desalination and Water Treatment 13, 109-119.

Volk, C., Renner, C., Joret, J.C., 1994. Comparison of two techniques for measuring biodegradable dissolved organic carbon in water. Environmental Technology 15, 545-556.

Vrouwenvelder, J.S., Bakker, S.M., Cauchard, M., Le Grand, R., Apacandie, M., Idrissi, M., Lagrave, S., Wessels, L.P., van Paassen, J.A., Kruithof, J.C., van Loosdrecht, M.C., 2007. The membrane fouling simulator: a suitable tool for prediction and characterisation of membrane fouling. Water Science and Technology 55(8-9), 197-205.

Weishaar, J.L., Aiken, G., Bergamaschi, B.A., Fram, M.S., Fujii, R., Mopper, K., 2003. Evaluation of specific ultraviolet absorbance as an indicator of the chemical composition and reactivity of dissolved organic carbon. Environmental Science and Technology 37, 4702-4708.

8

LOW MOLECULAR WEIGHT CUT-OFF ULTRAFILTRATION AS PRETREATMENT TO SEAWATER REVERSE OSMOSIS

It is not clear to what extent micro- and ultrafiltration (MF/UF) membranes reduce particulate/organic and biological fouling potential due to algal organic matter (AOM) during algal bloom periods to achieve the desired low cleaning frequency of SWRO membranes. The main goal of this chapter was to compare the effectiveness of 10 kDa and 150 kDa UF membranes in reducing organic and particulate fouling of SWRO feed water through removal of algal organic matter. 10 kDa and 150 kDa capillary UF membranes were examined and compared in terms of permeate quality and hydraulic performance. The membranes were tested with algal organic matter produced by laboratory cultivated cultures of the marine diatom *Chaetoceros affinis*. Permeate quality was assessed in terms of biopolymer concentration (LC-OCD), fluorescent compounds (F-EEM) and $MFI\text{-}UF_{10kDa}$. Results showed that 10 kDa membranes were capable of removing colloidal/organic fouling potential of RO feed water more effectively than 150 kDa membranes. 10 kDa completely removed AOM biopolymers and reduced fouling potential as measured by $MFI\text{-}UF_{10kDa}$ to below detection levels. In terms of hydraulic operation, the recovery of permeability with hydraulic backwash was less effective for 10 kDa than 150 kDa membranes. This difference in backwashing efficiency was attributed to the higher loading on the 10 kDa membranes due to higher retention and to the much lower surface porosity of the 10 kDa membranes revealed by FE-SEM imaging. The backwashability of the 10 kDa is expected to be largely improved by making membranes with surface porosity similar to the 150 kDa.

8.1 Introduction

Rapid growth in the market for seawater reverse osmosis (SWRO) has led to substantial reduction in the unit cost of these systems (Lee et al., 2011). Plant capacities have increased worldwide with extra large RO plants of 500,000 m^3/d production capacity constructed to meet increased water demand (Kurihara et al., 2013). A major concern for SWRO desalination is membrane fouling which affects process efficiency both in terms of quality and quantity of produced water (Matin et al., 2011) and results in longer downtime and higher operational costs. In severe cases, where fouling cannot be removed by chemical cleaning, membrane elements have to be replaced adding to the total cost of operation of the plant. It is obvious that operators prefer very low chemical cleaning frequency of the RO membranes to minimize the risk of membrane damage, downtime and quantities of cleaning solutions to be treated and discharged.

Several types of fouling are identified, i.e., biological fouling (biofouling), organic fouling, crystalline fouling (scaling) and particulate fouling (Al-Juboori et al., 2012; Matin et al., 2011).

Biofouling is a complex process caused by adhesion, accumulation and multiplication of microorganisms, mostly bacteria, on the membrane surface and/or feed spacers creating a densely concentrated sticky layer of bacteria and their extracellular secretions known as biofilm (Abd El Aleem et al., 1998). When biofilm accumulation leads to operational problems in RO membranes, biofouling occurs (Flemming, 2002). Biofouling may occur as membrane biofouling or as spacer/bundle clogging. Operational problems are manifested as flux decline or increase in net driving pressure, increase in feed channel pressure drop or a combination thereof. Biofouling eventually leads to biodegradation of cellulose acetate membranes resulting in increased salt passage.

Measuring the biofouling potential of RO feed water is rather complicated. A number of parameters have been proposed as indicators of biofouling potential. In practice, turbidity, total suspended solids (TSS) and silt density index (SDI) are commonly applied. However, no biological information can be obtained from these measurements. Liberman and Berman (2006) proposed a suite of parameters to determine the microbial support capacity (MSC) of RO feed water including parameters related to algal activity (e.g., chlorophyll-a, transparent exopolymer particles (TEP)), bacterial activity (e.g., adenosine trisphosphate (ATP)), total bacterial count, microscope observations, and concentration of growth-promoting nutrients (total N, total P).

Other biologically-oriented parameters include assimilable organic carbon (AOC) and biodegradable dissolved organic carbon (BDOC) (Vrouwenvelder and van der Kooij, 2001; Amy et al., 2011). These parameters are mainly applied in non-saline waters. Furthermore, several inline monitors such as the biofilm monitor and the membrane fouling simulator (MFS) were introduced to measure biofilm formation rate (Vrouwenvelder and van der Kooij, 2001; Vrouwendvelder et al., 2006). These monitors do not fully predict the biofouling potential of feed water but rather simulate the operation of a full scale plant. As such they can minimize costly pilot testing.

Organic fouling occurs when organic material accumulates on the membrane surface resulting in a decrease in normalized flux and/or increase in feed channel/bundle pressure drop. Organic matter may comprise algal and bacterial debris, algal organic matter (AOM) composed of high molecular weight biopolymers (polysaccharides and proteins), e.g., sticky TEP (Myklestad, 1995). Accumulation of sticky organic matter on the membrane surface may initiate particulate and biofouling by enhancing the deposition of bacteria and other particles from feed water to the membrane and spacers (Berman and Holenberg, 2005; Winters and Isquith, 1979). Advanced techniques for organic matter quantification and characterization, such as liquid chromatography - organic carbon detection (LC-OCD) for the detection of high molecular weight organic matter (Huber et al., 2011) and TEP measurements (Passow and Alldredge, 1995; Villacorte et al., 2013) can be used as indicators of the organic fouling potential of RO feed water. Organic matter might be present as suspended, colloidal and dissolved. It would be logical to categorize particulate organic matter under particulate fouling. However is it very difficult to exactly determine the category to which organic compounds in water belong.

Scaling occurs when the concentration of sparingly soluble inorganic compounds present in the feed water increases along the membrane module (brine flow) to such an extent that their solubility is exceeded and precipitation occurs. Antiscalants with and without acid are commonly used to control scaling in RO systems.

Particulate fouling is caused by the deposition of suspended and/or colloidal matter on the membrane surface, forming a heterogeneous cake/gel layer which may eventually cause substantial decline in overall membrane permeability. Suspended/colloidal matter may comprise inorganic particles, organic macromolecules, algae, algal detritus, bacteria, algal organic matter including TEP. Reverse osmosis membranes are primarily designed to remove dissolved constituents from the water and are highly vulnerable to spacer/bundle clogging problems by particulate matter and bacterial growth.

Two common methods are applied for measuring the particulate fouling potential of RO feed water, namely, the silt density index (SDI) and the Modified Fouling Index (MFI). SDI is the most widely used technique in SWRO plants. However, SDI has no reliable correlation with the concentration of particulate/colloidal matter (Alhadidi et al., 2013) and particles smaller than 0.45 μm are not accounted for. A more promising approach for measuring particulate fouling potential of RO feed water is the MFI. The index was developed by Schippers and Verdouw (1980) and adapted by Boerlage et al. (2004) and Salinas Rodriguez et al. (2009) for constant flux measurements with low molecular weight cut-off UF membranes (MFI-UF) to incorporate smaller colloidal particles. There is strong indication that these smaller colloids contribute significantly to particulate and organic fouling in RO systems (Salinas Rodriguez, 2011; Villacorte, 2014).

RO plants are equipped with pretreatment systems to ensure a chemical cleaning frequency which is acceptable and preferably convenient. This cleaning frequency is usually governed by the rule of thumb that - in spiral wound membrane systems - cleaning is recommended when increase in net driving pressure (NDP) and/or normalized pressure drop in the feed/concentrate channel exceeds 10 - 15 %.

Configuration and degree of required pretreatment depends on feed water quality and membrane type (Abd El Aleem, 1998). Pretreatment can be conventional including coagulation/flocculation, clarification (e.g., sedimentation or flotation) and/or granular media filtration (e.g., dual media filtration), or advanced using low pressure membrane technology (e.g., micro- and ultrafiltration MF/UF). In recent years, MF/UF has gained market as pretreatment for SWRO as they produce better quality feed water in terms of SDI and turbidity.

Algal blooms can enhance various forms of fouling in SWRO systems due to increased total suspended and colloidal material of organic nature resulting from algal biomass in the raw water (Caron et al., 2010). Therefore effective removal of algal organic matter (AOM) with pretreatment is desirable. Conventional coagulation relies on high coagulant dose (up to 20 mg Fe (III)/L) to substantially reduce AOM in SWRO feed water. UF membranes can substantially improve SWRO feed water quality without the need for coagulation. However, if higher removal of AOM is desired with currently applied UF membranes, coagulation at relatively high coagulant dose is needed. It is expected that the application of UF membranes with small pore size could further enhance AOM removal from SWRO feed water without relying on coagulation.

The main goal of this research was to compare the effectiveness of 10 kDa and 150 kDa UF membranes in reducing organic and particulate fouling potential of SWRO feed water through removal of algal organic matter. The specific objectives were,

- To evaluate the permeate quality of 10 kDa and 150 kDa UF membranes in terms of biopolymers concentration (measured by LC-OCD), fluorescent organics (measured by F-EEM) and constant flux MFI-UF_{10kDa} for feed waters containing laboratory cultivated marine algal species.
- To evaluate and compare the hydraulic performance of 10 kDa and 150 kDa UF membranes in terms of backwashability, treating feed water containing laboratory cultivated marine diatoms.

8.2 Particulate fouling prediction in RO

The time required for a defined net driving pressure increase (e.g., 15%) in RO systems, equipped with spiral wound elements - due to particle deposition on the membrane surface - can be predicted theoretically by using the cake/gel filtration model for measuring particulate fouling potential of RO feed water (Eq. 8-1). The model is based on the assumption that cake/gel filtration is the dominating mechanism in particulate fouling of RO membranes (Schippers et al., 1981) and does not incorporate other fouling mechanisms such as blocking or long term cake/gel compression.

$$t_r = \frac{(\Delta NDP)}{J^2.\eta.I.\Omega.\psi} \qquad \text{(Eq. 8-1)}$$

Where,

ΔNDP is increase in net driving pressure (bar) - typically cleaning criteria in RO systems

I is fouling index (m^{-2}), which can be derived from experimentally measured MFI at constant flux. MFI values depend on pore size of the membrane filters used and filtration flux (Salinas Rodriguez, 2011)

Ω is deposition factor and is measured on site or assumed

ψ is cake/gel ratio factor.

Particle deposition factor Ω represents the ratio of particles deposited on the RO membrane to those present in water passing the membrane (Schippers and Kostense, 1981). The value of Ω may range from 0 to 1 and has to be determined experimentally or assumed. Particle deposition factor is expected to depend on the flux, cross flow velocity and recovery in RO membranes, and the quality of the water to be tested. It is calcualted from the relation between the MFI of the concentrate ($MFI_{conc.}$) at recovery R, and the MFI of the feed water (MFI_{feed}). Assuming ψ=1, deposition factor is given by,

$$\Omega = \frac{1}{R} + \frac{MFI_{conc.}}{MFI_{feed}}\left(1 - \frac{1}{R}\right) \qquad \text{(Eq. 8-2)}$$

The cake/gel ratio factor ψ, accounts for differences between the cake deposited in MFI-UF measurements and the cake deposited on the RO membrane on short and long term. The cake/gel ratio factor ψ might deviate from 1, e.g., due to difference in deposition factor of particles with different size and characteristics resulting in a change in specific permeability of the cake/gel and prolonged compression of the cake/gel on the membranes surface during plant operation.

It has to be noted that the accuracy and reliability of such a prediction model (Eq. 8-1) for RO cleaning frequency based on particulate/organic fouling is limited by several aspects. Cake/gel ratio factor and deposition factor are empirical parameters whose values are generally not known and not easy to measure. Measurement of the average deposition factor in a plant or single element in a lab scale test unit is possible, under the assumption that ψ is 1. It requires accurate measurements and will depend on the choice of membrane pore size with which MFI measurements are carried out. A complicating factor is that the flux in a seawater RO system varies strongly, e.g., 30 to 5 L/m²h. Cake ratio factor ψ is very difficult to measure and quantify and cannot be assumed with confidence. Furthermore, the MFI or I values that are used as input for the prediction model are based on short term measurements and may not fully reflect the fouling phenomena (e.g., cake compression) that occur in an RO system in the long term.

Another shortcoming of the approach is the use of 10 kDa membranes for MFI measurement which does not fully reflect actual conditions in an RO system. Theoretically MFI measurements should be done with RO membranes. However, the use of RO membranes will cause strong interference of the measurements by osmotic pressure and scaling. Consequently membranes having a pore size as close as possible to RO membranes should be applied. However, these will take into account dissolved organic compounds as well. From a practical point of view, membranes with MWCO of 5 kDa, 1 kDa and ideally 0.5 kDa should be used to verify particles that really contribute to fouling of RO membranes. It cannot be excluded that very small particles will not contribute to fouling due to high permeability of the cake/gel formed and/or low deposition factor. Finally, the effect of cake/gel enhanced osmotic pressure is not taken into account. However, there are indications that this phenomenon has a substantial effect as well (Chong et al., 2008).

Nonetheless for the purpose of illustration the model serves as a tool to indicate - under certain assumptions - the extent to which different pretreatment methods affect cleaning frequency of RO systems.

8.3 Materials and methods

8.3.1 Membrane characterization

Membrane fibres with nominal MWCO of 10 kDa and 150 kDa were characterized for chemical composition, pore size distribution and surface porosity. Prior to characterization tests, membranes were cleaned by filtering Milli-Q water through the modules at 50 L/m^{2}h for 40 minutes. The modules were opened and fibres were removed and placed in a clean Petri dish and dried in a desiccator for 3 days at room temperature. The fibres were then cut longitudinally to expose the internal surface (active filtration layer), and flattened between two microscope glasses previously cleaned with ethanol. Care was taken to impart minimum damage to the fibres and the membrane surface.

8.1.1.1 Fourier transform infrared (FTIR) spectroscopy

FTIR was applied to identify the chemical composition and functional groups of the 10 kDa and 150 kDa UF membranes. Samples were analysed using a PerkinElmer ATR-FTIR Spectrum 100 instrument.

8.1.1.2 Field emission - scanning electron microscopy (FE-SEM)

FE-SEM images were taken of the active surface of 10 kDa and 150 kDa membrane samples to investigate surface porosity and pore size distribution of the two membranes. Voltage adjustment was required to avoid charging of samples. Samples were not sputtered with gold to avoid changes in surface morphology. FE-SEM images were processed with AutoCAD. Pores were identified based on visual judgement. Pore diameters were determined and data was manually transferred to an MS Excel sheet based on which pore size distribution (PSD) and membrane surface porosity were determined. Surface porosity was calculated as,

$$\text{Surface porosity } (\%) = \frac{\sum \text{pore area}(\text{nm}^2)}{\text{Total image area}(\text{nm}^2)}$$

PSD graph was obtained by categorising pores based on size and plotting frequency of pores of a certain size range by pore number or pore area.

8.3.2 Feed water preparation

Feed water was prepared by diluting stock AOM solutions - based on biopolymer concentration - obtained from the marine diatom species *Chaetoceros affinis*, (CCAP 1010/27) in synthetic seawater with a TDS of 35 g/L. Detailed procedure of AOM production and the recipe for synthetic seawater preparation are given in Chapter 7, section 7.2.1, 7.2.2 and 7.2.3. AOM was assumed to be mostly extracellular algal organic matter, as no cell destruction was attempted.

8.3.3 Water quality characterization

8.3.3.1 Liquid chromatography - organic carbon detection (LC-OCD)

Quantification and fractionation of dissolved organic carbon (DOC) was performed by the Liquid Chromatography - Organic Carbon Detection (LC-OCD) method of DOC-Labor (Karlsruhe, Germany). DOC was measured in the column bypass after inline filtration through 0.45 μm. For definitions and size ranges formally assigned (Huber et al., 2011) to different fractions of organic carbon by DOC-Labor refer to Chapter 7, section 7.2.6.2. The theoretical maximum size that can be analyzed by this method is 2 μm, resulting from the opening size in the frits at the entrance and exit of the SEC column that hold the packing material in place. The detection limit for DOC measurements ranges from 1 to 50 μg C/L and is compound specific.

8.3.3.2 Fluorescence excitation-emission matrix (F-EEM)

Fluorescence EEMs were obtained using a FluoroMax-3 spectrofluorometer (HORIBA Jobin Yvon, Inc., USA), with a 150 W ozone-free xenon arc-lamp as a light source for excitation and a 4 mL, 1 cm path length cuvette. Emission spectra were generated for each sample by scanning over excitation wavelengths between 240 and 450 nm at intervals of 10 nm and emission wavelengths between 290 and 500 nm at intervals of 2 nm. The slit widths on excitation and emission modes were both set at 1 nm. For details on experimental procedure and fluorescence responses of typical compounds (Leenheer and Croué, 2003) refer to Chapter 7, section 7.2.6.1.

Algal organic matter, including TEP, mainly comprises polysaccharides and glycoproteins. Polysaccharides do not fluoresce when excited by light. Protein-like organic matter exhibit a dominant peak at lower excitation/emission wavelengths while humic/fulvic substances show dominant primary and secondary peaks at higher excitation/emission wavelengths.

8.3.3.3 MFI-UF$_{10kDa}$

Flat sheet, regenerated cellulose (RC) membranes (Millipore, USA) of 25 mm diameter and a MWCO of 10 kDa were used for the experiments. The membranes were soaked in Milli-Q water to wet the pores and clean the membranes. Filtration flux for MFI-UF$_{10kDa}$ should resemble flux values in seawater RO membranes, i.e., 15 L/m^{2}h. This low flux could not be applied in the experimental setup for MFI-UF measurements, as the signal to noise ratio is very low at low flux values. Hence, MFI-UF$_{10kDa}$ was performed at a filtration flux of 60 L/m^{2}h. Filtration volume was approximately 60 mL. Transmembrane pressure (TMP) increase over time was recorded. Fouling index (I) was calculated based on the minimum slope of TMP versus time plots. MFI-UF was calculated by normalising I values to reference conditions for temperature (20 °C), pressure (2 bar) and membrane area (13.8 * 10^{-4} m^2). MFI-UF values were calculated from data normalised for 10 minutes moving average.

8.3.4 Filtration experiments

The performance of 10 kDa and 150 kDa capillary membranes in terms of fouling potential (MFI-UF) and backwashability was assessed using a laboratory scale filtration system operated in constant flux dead-end mode. The filtration system (Figure 8-1) consisted of a pen-sized UF module fed from both ends, while permeate was drawn from the centre of the module. Filtration was performed in inside-out mode. Backwashing was performed with UF permeate or synthetic seawater. Pressure was measured with a pressure sensor (PMC51, Endress + Hauser) and values of pressure and time were recorded in MS Excel using a communication device (FXA 195 Hart, Endress + Hauser) and data acquisition software (OPC Office Link, Rensen).

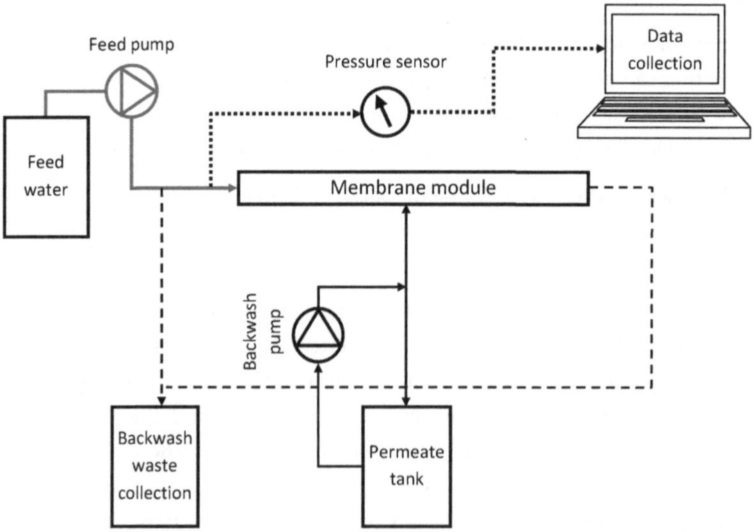

Figure 8-1 Scheme of experimental for filtration experiments

Membrane modules were prepared by potting 6 polyethersulphone (PES) UF capillaries with nominal MWCO of 10 kDa or 150 kDa and internal diameter of 0.8 mm (X-Flow, Pentair) in 30 cm transparent flexible polyethylene tubing (PEN-x1, 25-NT, Festo). Membrane modules were potted using polyurethane glue. Effective membrane area was approximately 0.0045 $m^2 \pm 5\%$. Operational conditions for the UF membranes are summarized in Table 8-1.

Table 8-1 Operating conditions for hydraulic experiments

Parameter	Value
Feed concentration	0.5 mg C/L as biopolymers
Filtration flux	50 L/m^2h
Number of cycles	15 cycles
Duration of one filtration cycle	20 minutes
Backwash flux	125 L/m^2h
Duration of one backwash cycle	45 seconds
Backwash water volume	10% of the total filtered volume in one cycle

Membrane fibres were cleaned rigorously to remove any preserving material and loose polymers of the membrane itself and prevent the contribution of these organic polymers to permeate samples. Cleaning procedure is described in Annex 1. Membrane resistance (R_m) was measured by filtering Milli-Q water through the membrane at the same flux as subsequent feed solutions for 40 minutes. Thereafter pre-filtered synthetic seawater (through 10 kDa) was filtered to condition the membranes for the same ionic conditions as subsequent feed solutions. Feed solutions were filtered through the membrane at 50 or 80 L/m^2h for 20 minutes. At the end of each filtration cycle hydraulic cleaning was performed by backwashing at 2.5 times the filtration flux for 45 seconds. These steps were repeated for 15 continuous cycles of filtration and backwash. Membrane performance was characterized in terms of pressure increase during a filtration cycle (bar/h) and increase in resistance due to non-backwashable fouling through multiple filtration cycles (m^{-1}).

8.4 Results and discussion

8.4.1 Membrane characterization

To explain the hydraulic performance of 10 kDa and 150 kDa membranes and the effectiveness of these membranes in reducing fouling potential of RO feed water, membranes were analyzed for chemical composition with FTIR and for surface porosity and pore size distribution by analysis of FE-SEM images.

8.4.1.1 FTIR spectra

FTIR analysis was performed on 10 kDa and 150 kDa UF membrane fibres as received from the manufacturer (virgin), and pre-cleaned with Milli-Q. Results are presented in Figure 8-2.

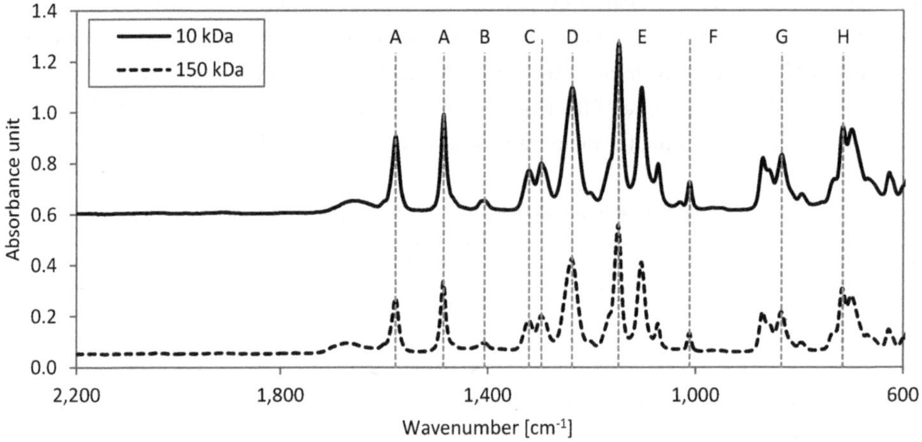

Figure 8-2 FTIR spectra of 10 kDa and 150 kDa membranes cleaned with Milli-Q water; offset in absorbance values were made for clarity

High absorbance values were observed at similar wavelengths for 10 kDa and 150 kDa membranes, indicating similar chemical composition. The functional groups identified based on the IR spectra are summarized in Table 8-2. The presence of typical bands at 1578 and 1485 cm^{-1} are assigned to aromatic C=C from the aryl ring and is in accordance with literature for PES membranes (Picot et al., 2012). From FTIR spectra, a difference in chemical composition between the two membranes could not be established.

Table 8-2 Absorption bands identified in the infrared spectra of 10 kDa and 150 kDa PES membranes

	Functional groups	Wavelength
A	Aromatic skeletal and asymmetric stretching vibrations	1,578 & 1,485
B	SO2 groups	1,406
C	Asymmetric O=S=O stretching of the sulfone group	1,321 & 1,297
D	Asymmetric and symmetric vibrations of C-O-C groups of alkyl aryl ethers	1,238
E	Symmetric O=S=O stretching of the sulfone group	1,148
F	Ring vibrations of para-substituted phenyl ethers	1,011
G	C-H	835
H	C-S	717

8.1.1.3 Visual observation of membrane surfaces

FE-SEM images of cleaned and un-cleaned 10 kDa and 150 kDa membranes are presented in Figure 8-3. The 10 kDa membranes did not show a significant difference before and after cleaning. However, for 150 kDa membranes a large number of pores of the virgin membranes were covered with preserving material and were not visible. Cleaning with Milli-Q had a large effect on the visual presence of pores and a significant number of pores were observed for 150 kDa after cleaning. FE-SEM images of cleaned membrane fibres were used for analysis of pore size distribution and surface porosity.

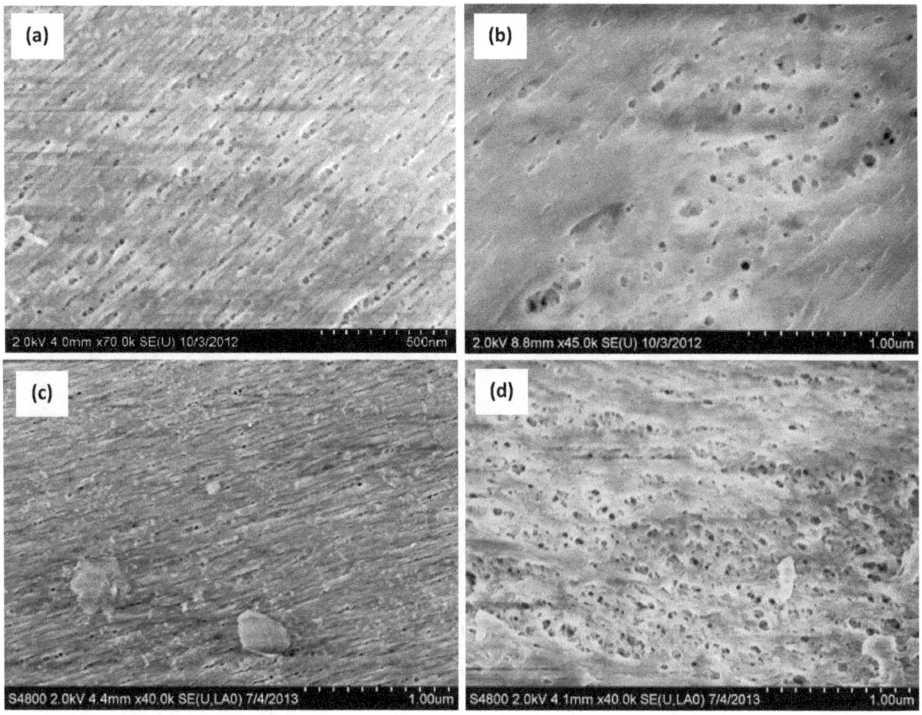

Figure 8-3 FE-SEM images of the surface of virgin (a & b) and cleaned (c & d) fibres of 10 kDa (left images) and 150 kDa (right images)

8.1.1.4 Pore size distribution and surface porosity

High magnification (~ x100,000) FE-SEM images were used for analysis of pore size distribution and surface porosity of 10 kDa and 150 kDa membranes (Figure 8-4).

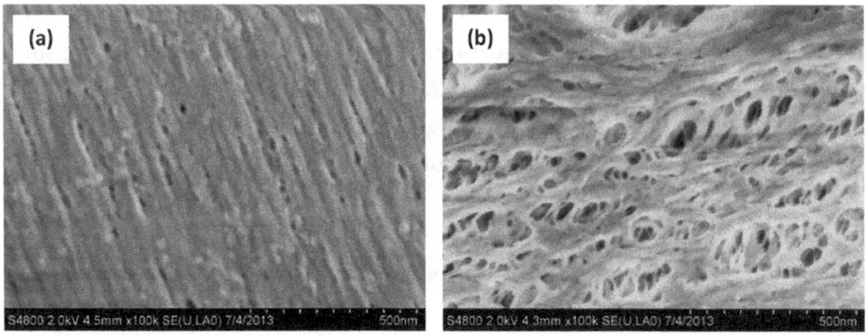

Figure 8-4 FE-SEM images of the surface of (a) 10 kDa and (b) 150 kDa membranes at 100,000x magnification

Pore size distribution (PSD) of 10 kDa and 150 kDa membranes normalized for a surface area of 1 μm^2 is shown in Figure 8-5. 10 kDa membranes showed a narrow PSD with most of the pores within the range of 6 to 18 nm. For 150 kDa membranes, PSD was wider with a substantially higher number of larger pores.

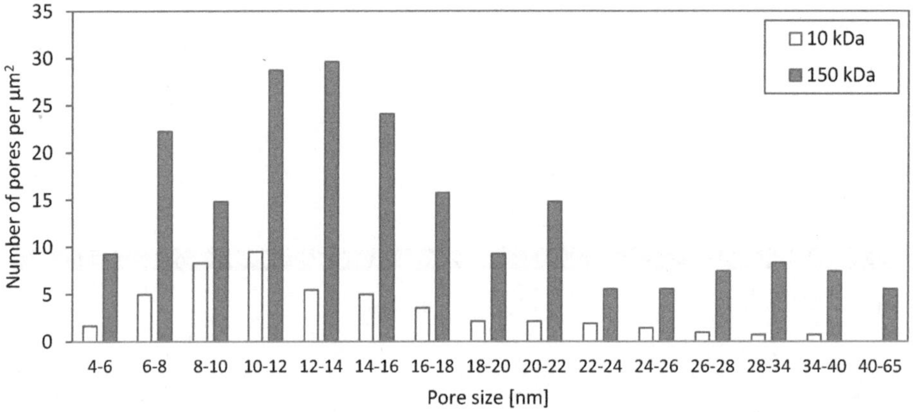

Figure 8-5 Pore size distribution of surface pores of 10 kDa and 150 kDa membranes, obtained from FE-SEM images

Table 8-3 shows differences between the 10 kDa and 150 kDa membranes in terms of surface porosity. Surface porosity of the 150 kDa UF membranes was significantly higher than that of 10 kDa membranes, i.e., about 7 times.

Table 8-3 Surface porosity of 10 kDa and 150 kDa membranes

Membrane type	Surface porosity (%)
10 kDa	0.9
150 kDa	6.3

Note: pore size measured with FE-SEM is not consistent with pore size given by the manufacturer based on molecular weight cut-off. Therefore, it is very likely that the surface pore size observed for 10 kDa membranes in FE-SEM images is larger than the actual pore size of these membranes. FE-SEM images give an impression of the pore size at the surface and not the real pore size. Measurement of MWCO gives a more realistic indication.

8.4.2 Hydraulic performance

The rate of fouling of 10 kDa and 150 kDa membranes by AOM solutions in synthetic seawater was assessed for different conditions. Filtration was performed at 50 L/m^2h for 30 minutes and permeate was used for backwashing at 125 L/m^2h for 1 minute. Development of TMP versus time and non-backwashable resistance at the start of each filtration cycle is shown in Figure 8-6.

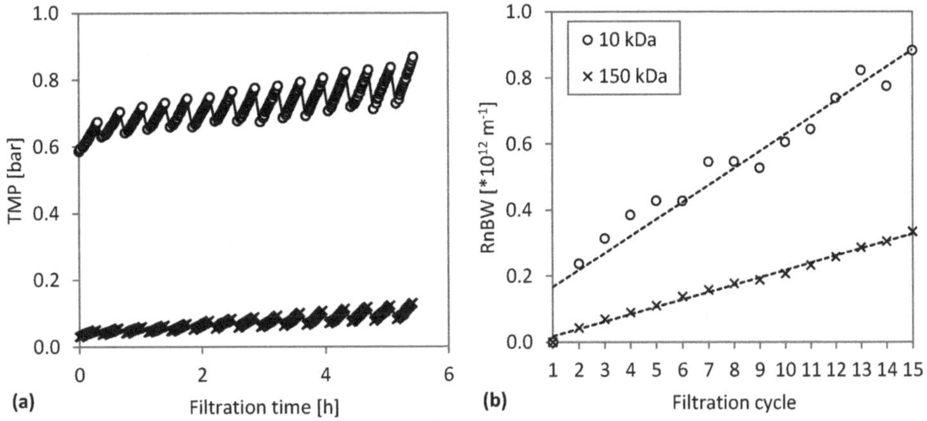

Figure 8-6 (a) TMP and (b) resistance development in time for filtration of AOM obtained from *Chaetoceros affinis* at 0.5 mg C/L as biopolymers through 10 kDa and 150 kDa UF membranes

Table 8-4 shows calculated values of MFI-UF, non-backwashable and total fouling rate for the two membranes. Fouling potential measured as MFI-UF was considerably higher for 10 kDa membranes; almost 4 times. This is attributed to the smaller pore size of the 10 kDa membranes as compared to 150 kDa membranes. Consequently, this type of membranes has to deal with a much larger loading of fouling components.

Table 8-4 MFI-UF, non-backwashable and total fouling rate for 10 kDa and 150 kDa membranes fed with AOM from *Chaetoceros affinis* at 0.5 mg C/L as biopolymers

MWCO	MFI-UF [s/L^2]	R_{nBW} [m^{-1}/h]	R_t [m^{-1}/h]
10 kDa	26,000	4.6 *10^{10}	7.4 *10^{10}
150 kDa	7,200	2.2 *10^{10}	3.8 *10^{10}

Rate of non-backwashable and total fouling of 10 kDa membranes was approximately two times that of 150 kDa membranes. This difference was attributed to the lower surface porosity of the 10 kDa membranes. During backwashing cake/gel layer is expected to be lifted fully or partly. During backwashing of the 10 kDa membranes water flows at high velocity through the pores. However, the porosity is too low for fully lifting the cake/gel layer from the membrane surface. Rather the cake/gel layer is partially lifted around the pores and if not carried out of the lumen by the cross flow might re-collapse on the membrane. For 150 kDa membranes, surface porosity is substantially higher. Therefore during backwashing, the cake/gel layer is lifted from the membrane surface more effectively, resulting in better backwash efficiency.

Filtration tests with 10 kDa and 150 kDa membranes showed a relation between average slope per cycle (bar/h) and non-backwashable resistance at the start of each cycle, for multiple filtration cycles (Figure 8-7). This relation suggests that a part of the membrane surface is increasingly blocked, resulting in a higher flux at the unblocked part of the membrane surface.

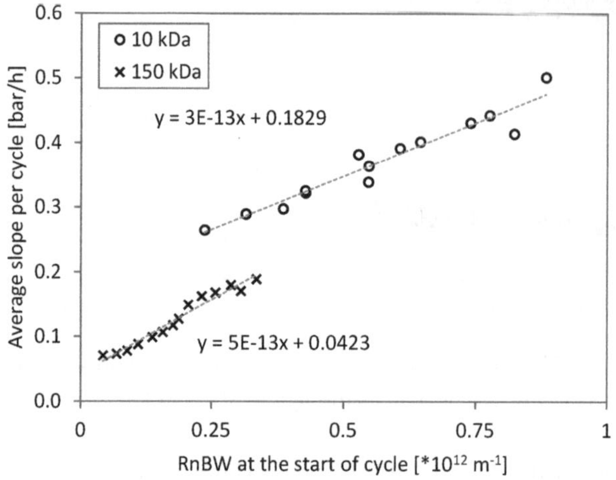

Figure 8-7 Correlation of average slope per filtration cycle and non-backwashable resistance at the start of each cycle for 15 successive cycles with 10 kDa and 150 kDa membranes at 50 L/m²h

There are indications that the quality and matrix of backwash water might affect UF operation (Li et al., 2011). It was hypothesized that backwashing with synthetic seawater enhances UF performance as UF permeate might contain organic matter which could block pores from the permeate side of the capillaries. Figure 8-8 shows the resistance profiles of 10 kDa and 150 kDa membranes backwashed with permeate of the respective membranes or synthetic seawater.

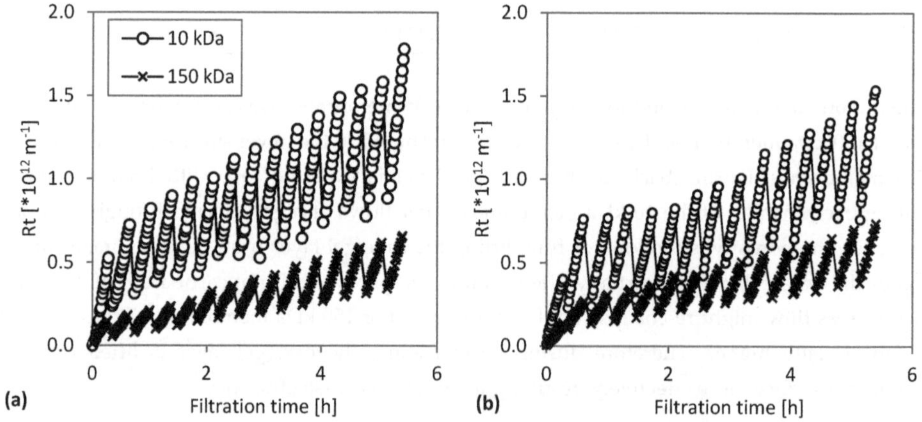

Figure 8-8 Resistance development due to fouling by AOM obtained from *Chaetoceros affinis* at 0.5 mg C/L as biopolymers for backwashing with (a) permeate and (b) synthetic seawater

Substantial differences in hydraulic performance of the two membranes could not be observed for backwashing with UF permeate or synthetic seawater. This rules out the possibility of membrane fouling from the permeate side due to the presence of biopolymers in permeate water.

8.4.3 Permeate quality

Permeate quality of 10 kDa and 150 kDa UF membranes was characterized to assess the effectiveness of the membranes in reducing particulate and organic fouling potential by measuring biopolymer concentration, fluorescent compounds and $MFI\text{-}UF_{10kDa}$. Based on MFI-UF values, cleaning frequency of RO membranes fed with permeate of 10 kDa and 150 kDa was projected.

8.4.3.1 DOC fractionation

LC-OCD chromatograms of feed water and permeate samples of 10 kDa and 150 kDa membranes are presented in Figure 8-9 for the filtration of AOM obtained from *Chaetoceros affinis*. Elution time of biopolymers in AOM harvested from *Chaetoceros affinis* is approximately 60 minutes. Biopolymer concentration in the permeate of 10 kDa membranes was below detection limit. This indicates that 10 kDa membranes are capable of completely removing biopolymers, for feed biopolymer concentration of 0.5 mg C/L. The biopolymer concentration in 150 kDa permeate was approximately 200 µg C/L. The difference in retention is attributed to the difference in pore size (MWCO) of the two membranes.

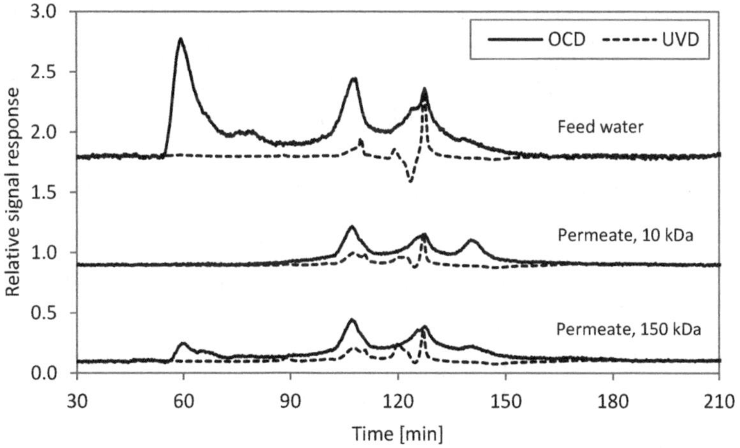

Figure 8-9 OCD and UVD chromatograms of feed water, and permeate of 10 kDa and 150 kDa membranes

8.4.3.2 Fluorescence spectra

F-EEM spectra of feed and permeate samples obtained from 10 kDa and 150 kDa membranes were analysed to verify the presence and removal of fluorescent organic substances (Figure 8-10). Fluorescence responses mainly originated from tryptophan-like proteins and no humic-like and fulvic-like peaks were observed in the feed sample. Protein-like peaks were reduced to the level of Milli-Q water, i.e., 0.02 - 0.03 RU, by both 10 kDa and 150 kDa membranes.

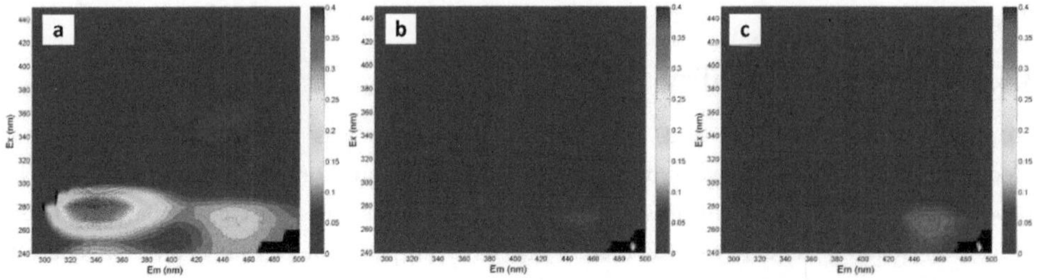

Figure 8-10 Fluorescence spectra of (a) feed water, (b) 10 kDa permeate and (c) 150 kDa permeate

8.4.3.3 MFI-UF$_{10kDa}$

MFI-UF$_{10kDa}$ was measured for permeate samples of 10 kDa and 150 kDa UF membranes for feed biopolymer concentrations of 0.5, 1 and 1.5 mg C/L. MFI-UF$_{10kDa}$ was measured at three different flux values for each sample, namely, 60 L/m^{2}h, 80 L/m^{2}h and 100 L/m^{2}h. Measurements were conducted using 10 kDa regenerated cellulose membranes. Results are shown in Table 8-5.

MFI-UF$_{10kDa}$ values of permeate samples of 10 kDa UF membranes were below detection limit irrespective of biopolymer concentration in the feed solution. This was attributed to high biopolymer removal from the feed solutions by 10 kDa membranes. MFI-UF$_{10kDa}$ was measurable in permeate samples of 150 kDa membranes and was higher at higher biopolymer concentration. This might indicate that biopolymer removal with 150 kDa depends on the biopolymer concentration in the feed. For these samples, higher MFI-UF$_{10kDa}$ was observed at higher filtration flux. This is in concert with observations of Salinas Rodriguez (2011).

Table 8-5 Measured MFI-UF$_{10kDa}$ values for permeate samples obtained from 10 kDa and 150 kDa

Feed biopolymer conc.	10 kDa			150 kDa		
	60 L/m^{2}h	80 L/m^{2}h	100 L/m^{2}h	60 L/m^{2}h	80 L/m^{2}h	100 L/m^{2}h
0.5 mg C/L	bdl	bdl	n.m	800	1500	5500
1.0 mg C/L	bdl	bdl	n.m	5200	10200	11500
1.5 mg C/L	bdl	bdl	n.m	9200	12600	18900

bdl = below detection limit (50 s/L^2); n.m = not measured

MFI-UF measurements should ideally be done using membranes with pores close to nanofiltration range to fully incorporate colloidal particles. Consequently, 10 kDa UF membranes were used for MFI-UF measurements. MFI-UF measurements with membranes having lower molecular weight cut-off of 0.5 kDa, 1 kDa and 5 kDa may be more suitable to measure colloidal/organic fouling potential of seawater, but these methods are not yet fully validated.

8.4.4 Prediction of cleaning frequency in RO membranes

Cleaning frequency in SWRO membranes was predicted based on $MFI\text{-}UF_{10kDa}$ values of 150 kDa permeate samples - measured at 60 - 100 L/m²h - using the model described in 8.2. Average flux in SWRO membranes is approximately 15 L/m²h. $MFI\text{-}UF_{10kDa}$ could not be measured at this low flux. Therefore projections were done by extrapolation to obtain $MFI\text{-}UF_{10kDa}$ at 20 L/m²h. Projections were made by,

- Simple extrapolation of the $MFI\text{-}UF_{10kDa}$ values to 20 L/m²h;
- Extrapolation by setting the Y-intercept to zero.

Projected values of $MFI\text{-}UF_{10kDa}$ at 20 L/m²h, using different extrapolation methods is shown in Table 8-6. R^2 values are higher for simple extrapolation. However, for predicting SWRO cleaning frequency we used values obtained through extrapolation by setting Y-intercept to zero. The real values of $MFI\text{-}UF_{10kDa}$ at 20 L/m²h, could lie anywhere outside or in between the extrapolated results.

Table 8-6 Projected MFI-UF values for permeate of 150 kDa at 20 L/m²h flux for feed solutions with different biopolymers concentration

	$MFI\text{-}UF_{10kDa}$ at 20L/m²h			
Sample	Extrapolation	R^2	Y-int. zero	R^2
150 kDa permeate (0.5 mg C/L)	0	1.00	1300	0.53
150 kDa permeate (1.0 mg C/L)	0	0.96	2200	0.55
150 kDa permeate (1.5 mg C/L)	0	0.92	3100	0.70

To predict cleaning frequency in RO membranes the following assumptions were made,

- Deposition factor of 1 (full deposition of biopolymers on the surface of RO membranes) and cake/gel ratio factor of 1 (similar RO and MFI-UF cake characteristics); therefore the product $\Omega * \psi = 1$. Measuring ψ will be very difficult. Salinas Rodriguez found average values for Ω in the range of 0.2 - 0.4 for MFI 100 kDa, 30 kDa and 10 kDa in a seawater RO pilot plant, assuming $\psi = 1$.
- Increase in net driving pressure (ΔNDP) 1 bar and 3 bar
- Flux 20 L/m²h.

Note: Accurate and reliable prediction of the rate of fouling and cleaning frequency of seawater RO plants based on MFI is not yet possible due to gaps in validated tools (MFI 0.5 kDa, 1 kDa and 5 kDa, and cake/gel enhanced osmotic pressure), and lack of data from full scale and pilot plants. To improve the current situation Schippers et al. (2014) recommend,

- Measuring MFI 10 kDa, 5 kDa, 1 kDa and 0.5 kDa of feed water and determining the product of deposition factor and cake/gel ratio factor ($\Omega * \psi$) in a multitude of full scale plants and to correlate this with the rate of fouling. Differentiation between particulate fouling and biofouling is required and a complicating factor.

- Measuring the effect of enhanced osmotic pressure due to fouling in full scale plants.

Figure 8-11 shows predicted cleaning frequency in months as a function of MFI for two target values of ΔNDP, i.e., 1 bar and 3 bars. In general, the higher the MFI, the higher the cleaning frequency (shorter time interval between cleanings). The set target for increase in net driving pressure also affects cleaning frequency; the higher the target increase in net driving pressure, the lower the cleaning frequency.

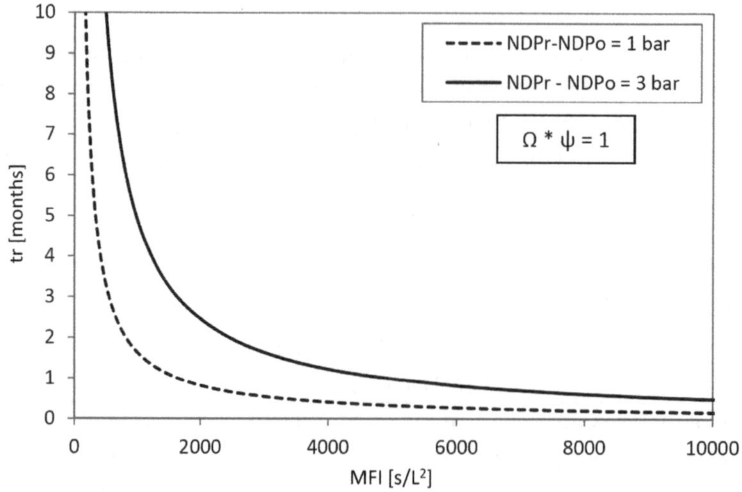

Figure 8-11 Cleaning frequency in months as a function of MFI; NDP_r - NDP_o = ΔNDP (adapted from Salinas Rodriguez, 2011)

It may be seen that cleaning frequency of RO membranes fed with permeate of 10 kDa membranes (MFI-UF_{10kDa} below detection limit, i.e., < 50 s/L^2) will be low with a predicted cleaning frequency of approximately 30 months for target ΔNDP of 1 bar. For RO membranes fed with permeate of 150 kDa membranes predicted cleaning frequency would be in the range of < 1 to 2 months for different feed biopolymer concentrations at target ΔNDP of 1 bar. As discussed earlier, the assumptions in these calculations make it difficult to make accurate predictions of cleaning frequency of SWRO membranes fitted with UF pretreatment systems. The reported cleaning frequencies are therefore only indicative.

8.5 Conclusions and recommendation

Based on FTIR spectra, no difference in chemical composition between the two UF membranes was noticed as similar absorbance profiles were obtained. Substantial differences in surface porosity (revealed by FE-SEM imaging) of the membranes were noted as surface porosity of 10 kDa and 150 kDa membranes was approximately 0.9% and 6% respectively.

Under identical operating conditions - treating AOM solutions in synthetic seawater - pressure development in successive filtration cycles was lower for 150 kDa membranes compared to 10 kDa membranes. This difference was attributed to firstly, fouling potential measured as MFI-UF was 7 times higher for 10 kda as compared to 150 kDa membranes and consequently 10 kDa had to deal with much higher loading of foulants. Secondly, the lower surface porosity of 10 kDa membranes resulted in lower backwashing efficiency. The backwashability of the 10 kDa is expected to be largely improved by making membranes with surface porosity similar to the 150 kDa.

No difference was observed in backwash efficiency of 10 kDa and 150 kDa between backwashing with UF permeate or synthetic seawater.

Membranes with a pore size of 150 kDa were capable of partly removing AOM biopolymers and fouling potential as measured by MFI-UF_{10kDa}. However, 10 kDa membranes completely removed AOM biopolymers and fouling potential as measured by MFI-UF_{10kDa} to below detection level. This difference in performance is attributed to the smaller pore size (MWCO) of the 10 kDa membranes. Thus producing better quality RO feed water in terms of particulate/organic fouling potential and reducing fouling problems in RO is expected.

8.6 Acknowledgements

Thanks are due to Mohaned Sousi, Sergio Salinas, DOC-Labor for LC-OCD analyses, CCAP for providing the diatom strain, Hozan Miro from TNW, TU Delft for FE-SEM analyses.

8.7 References

Abd El Aleem FA, Al-Sugair KA, Alahmad MI (1998) Biofouling problems in membrane processes for water desalination and reuse in Saudi Arabia. International Biodeterioration and Biodegradation 41: 19-23.

Alhadidi, A., Blankert, B., Kemperman, A. J. B., Schurer, R., Schippers, J. C., Wessling, M., & van der Meer, W. G. J. (2013). Limitations, improvements and alternatives of the silt density index. Desalination and Water Treatment, 51(4-6), 1104-1113.

Al-Juboori RA, Yusaf T (2012) Biofouling in RO system: Mechanisms, monitoring and controlling. Desalination 302: 1-23 DOI 10.1016/j.desal.2012.06.016

Amy, G.L., Salinas-Rodriguez, S.G., Kennedy MD, Schippers JC, Rapenne S, Remize P-J, Barbe C, Manes CLDO, West NJ, Lebaron P, Kooij DVD, Veenendaal H, Schaule G, Petrowski K, Huber S, Sim LN, Ye, Y., Chen, V., Fane, A.G., 2011. Water quality assessment tools. In: Drioli, E., Criscuoli, A. & Macedonio, F. (eds.) Membrane-Based Desalination - An Integrated Approach (MEDINA). IWA Publishing, New York, pp 3-32.

Berman T (2010) Biofouling: TEP – a major challenge for water filtration. Filtration & Separation 47, 20-22.

Berman T, Holenberg M (2005) Don't fall foul of biofilm through high TEP levels. Filtration & Separation 42, 30-32.

Boerlage, S.F.E., Kennedy, M., Tarawneh, Z., de Faber, R., Schippers, J.C., 2004. Development of the MFI-UF in constant flux filtration. Desalination. 161, 103-113.

Caron , D.A., Garneau, M.E., Seubert, E., Howard M.D.A.,, Darjany L., Schnetzer A., Cetinic, I., Filteau, G., Lauri, P., Jones, B. and Trussell, S. (2010), "Harmful Algae and Their Potential Impacts on Desalination Operations of Southern California", Water Research, 44, pp. 385–416.

Chong, T.H., Wong, F.S., Fane, A.G., 2008. Implications of critical flux and cake enhanced osmotic pressure (CEOP) on colloidal fouling in reverse osmosis: Experimental observations. Journal of Membrane Science 314(1-2), 101-111.

Flemming, H. C. (2002) Biofouling in water systems–cases, causes and countermeasures. Applied Microbiology and Biotechnology 59, 629-640.

Huber S.A., Balz A., Abert M. and Pronk W. (2011) Characterisation of aquatic humic and non-humic matter with size-exclusion chromatography – organic carbon detection – organic nitrogen detection (LC-OCD-OND). Water Research 45:879-885.

Kurihara M, Hanakawa M (2013) Mega-ton Water System: Japanese national research and development project on seawater desalination and wastewater reclamation. Desalination 308, 131-137.

Lee, K.P., Arnot, T.C., Mattias, D., 2011. A review of reverse osmosis membrane materials for desalination—Development to date and future potential. Journal of Membrane Science 370, 1-22.

Leenheer, J.A., Croue, J.P., 2003. Characterising aquatic dissolved organic matter. Environmental Science and Technology 37 (1), 18A-26A.

Li, S., Heijman, S.G.J., Verberk, J.Q.J.C., Le Clech, P., Lu, J., Kemperman, A.J.B., Amy, G.L., van Dijk, J.C., 2011. Fouling control mechanisms of demineralised water backwash: Reduction of charge screening and calcium bridging effects. Water Research, 45 (19), 6289-6300.

Liberman B, Berman T (2006) Analysis and monitoring: MSC – a biologically oriented approach. Filtration & Separation 43, 39-40.

Matin A, Khan Z, Zaidi SMJ, Boyce MC (2011) Biofouling in reverse osmosis membranes for seawater desalination: Phenomena and prevention. Desalination 281, 1-16.

Myklestad, S. M. (1995). Release of extracellular products by phytoplankton with special emphasis on polysaccharides. Science of the Total Environment, 165, 155-164.

Passow and Alldredge, 1995 Passow, U. and Alldredge, A. L. (1995) A Dye-Binding Assay for the Spectrophotometric Measurement of Transparent Exopolymer Particles (TEP). Limnology and Oceanography 40(7), 1326-1335.

Picot, M., Rodulfo, R., Nicolas, I., Szymczyk, A., Barrière, F., Rabiller-Baudry, M., 2012. A versatile route to modify polyethersulfone membranes by chemical reduction of aryldiazonium salts. Journal of Membrane Science 417–418, 131-136.

Salinas Rodríguez S. G., Kennedy M. D., Schippers J. C., Amy G. L., 2009. Organic foulants in estuarine and bay sources for seawater reverse osmosis – Comparing pretreatment processes with respect to foulant reductions. Desalination and Water Treatment 9 (1-3), 155-164.

Salinas Rodriguez, S.G., 2011. Particulate and organic matter fouling of SWRO systems: characterization, modelling and applications. PhD thesis, UNESCO-IHE/TU Delft, Delft.

Schippers, J.C., and Verdouw, J., 1980. The modified fouling index, a method of determining the fouling characteristics of water. Desalination. 32, 137-148.

Schippers, J.C., Kostense, A., 1981. The effect of pretreatment of River Rhine on fouling of spiral wound reverse osmosis membranes. In: Proceedings of the 7th International Symposium on Fresh Water from the Sea, Amsterdam, 297-306.

Schippers, J.C., Salinas Rodriguez, S.G., Boerlage, S., Kennedy, M.D., 2014. History and development of the MFI and SDI. IDA Journal of Desalination and Water Reuse. In Press.

Villacorte, L.O., 2014. Algal blooms and membrane based desalination technology. Doctoral dissertation, UNESCO-IHE/TUDelft, Delft.

Villacorte, L.O., Ekowati, Y., Winters, H., Amy, G.L., Schippers, J.C. and Kennedy, M.D. (2013) Characterisation of transparent exopolymer particles (TEP) produced during algal bloom: a membrane treatment perspective. Desalination & Water Treatment 51 (4-6), 1021-1033.

Vrouwenvelder JS, van der Kooij, D., 2001. Diagnosis, prediction and prevention of biofouling of NF and RO membranes. Desalination 139, 65-71.

Vrouwenvelder, J. S., van Paassen, J. A. M., Wessels, L. P., van Dam, A. F., & Bakker, S. M. (2006). The membrane fouling simulator: A practical tool for fouling prediction and control. Journal of Membrane Science, 281(1-2), 316-324.

Winters, H., and Isquith, I. R. (1979). In-plant microfouling in desalination. Desalination, 30(1), 387-399.

Annex 1

Cleaning protocol for UF membranes

1. Place the membrane module in the setup and remove air by flushing the inside of the fibres from the feed side. Filter Milli-Q water through the membranes at 120 L/m^2h. Preservative materials will be removed partially from the membrane surface and pores during this step.
2. Filter Milli-Q through the membranes at 50 L/m^2h for about 40 minutes (record pressure and time during filtration to ensure membrane integrity) to increase the removal of the preservative materials (minimum volume of 200 mL Milli-Q water).
3. Remove module from the filtration setup and soak it in a bucket filled with 4 litres of 30 ^oC Milli-Q water for 24 hours in a temperature controlled room. Frequently shake the bucket containing the module throughout the soaking period.
4. After 12 hours of soaking, replace the water with fresh warm water.
5. After 24 hours of soaking, place the module in the filtration setup again and flush the fibres with Milli-Q water at 50 L/m^2h through the fibres for 40 minutes (record the pressure versus time again during the MQ water filtration to check whether the membrane is damaged during soaking or not by comparing the membrane resistance before and after soaking).
6. Filter 140 ml of synthetic seawater through the fibres to condition the membranes to matrix of feed water in subsequent filtration experiments and determine membrane resistance.

9

GENERAL CONCLUSIONS

9.1 Conclusions

In seawater desalination, operation of reverse osmosis (RO) membranes and pretreatment systems is challenging during periods of algal bloom where relatively high concentrations of algal cells and algal organic matter (AOM) are present in the feed water. During such periods, pretreatment systems have to be capable of operating smoothly without extensive clogging and/or fouling to ensure constant production of feed water for seawater reverse osmosis (SWRO). In addition, pretreatment systems have to provide high quality feed water to prevent particulate/organic and biofouling in SWRO membranes.

Operational experience during a severe algal bloom in the Middle East showed that granular media filtration (GMF) - in combination with coagulation - was incapable of reliable operation in such circumstances, both in terms of clogging of the media filters and SWRO feed water quality. Limited published data on operational experience with ultrafiltration (UF) as pretreatment to SWRO during algal bloom periods has shown that UF membranes can be operated on algal bloom impaired water with high fouling propensity. However, coagulation was required to stabilize UF operation.

In general, UF membranes produce better SWRO feed water quality than GMF in terms of turbidity and silt density index (SDI). However, during algal bloom periods, the removal of AOM from SWRO feed water is of interest. AOM can initiate and exacerbate organic/particulate and biofouling and associated problems in SWRO membranes. This aspect is often overlooked as biofouling is a slow process and may not be observed for a long time after the bloom has subsided. Coagulant addition can enhance the removal of AOM by UF membranes.

Coagulation entails process complexities and costs associated with chemicals (coagulants, coagulant aids, pH adjustment), dosing equipment, storage capacity for chemicals, automation, etc. Moreover, if backwash water containing coagulant needs to be treated before discharge, higher investment and operational costs may be incurred for treatment, sludge handling and disposal. Given the complexities and costs associated with coagulation in UF/RO systems for seawater desalination, complete elimination of coagulation from these systems is highly desired.

This study looked at the role of coagulation in UF operation as pretreatment to SWRO in terms of hydraulic performance and AOM removal during algal bloom periods, with a particular emphasis on minimizing or ideally eliminating coagulants from the process scheme. The following approach was taken,

- Elucidating the effect of coagulation conditions on UF operation
- Investigating alternative modes of coagulant application (e.g., coating of UF membranes) that require minimal or no dose of metal-based coagulant to stabilize UF hydraulic performance

- Identifying alternative membrane properties (e.g., pore size, surface porosity, etc.) that require less or no coagulant to remove AOM from RO feed water

The overall aim of this research was to provide insight into technical options for coagulant free or very low coagulant consuming UF operation on seawater for the production of high quality SWRO feed water.

9.1.1 Role of coagulation in MF/UF systems

Micro and ultrafiltration (MF/UF) systems rely on coagulation - in inline mode - mainly to stabilize hydraulic performance. However, no guidelines have been established for inline coagulation prior to MF/UF systems. The mechanisms with which coagulation enhances MF/UF operation are attributed to reduced pore blocking by enlarging particle size and enhanced permeability by increased cake/gel layer porosity.

Using a simple model based on the Carman-Kozeny equation, the role of particle size, density and cake/gel layer porosity on pressure increase in MF/UF membranes in one filtration cycle was investigated. Calculations showed that the contribution of particle properties such as size and density, and porosity of the cake/gel layer to pressure increase in MF/UF systems in one filtration cycle is marginal. For particle sizes ranging from a few micrometres down to a few nanometres, pressure development in one filtration cycle was well below threshold levels for hydraulic backwash (0.4 bar increase in pressure).

Furthermore, the effect of different process conditions (pH, G, Gt and coagulant dose) on inline coagulation pretreatment of MF/UF with ferric chloride was studied. Inline coagulation mainly reduced the fouling potential of different feed waters, resulting in low predicted pressure increase in MF/UF systems. Flocculation demonstrated marginal improvement. Results indicated that absence of flocculation at moderate G-values can be compensated by applying higher coagulant dose. Calculations showed that the high G-values and short residence times available prior to and within MF/UF elements are sufficient to maintain/achieve relatively low fouling potential of the feed water arriving at the membrane surface. The results of this chapter suggest that the main mechanism involved in fouling control in inline coagulation is most likely the improvement of backwashability of the fouling layer.

9.1.2 Coagulation/UF of algal bloom-impaired seawater

The effect of coagulation on hydraulic performance and permeate quality of UF membranes fed with solutions of AOM (obtained from *Chaetoceros affinis*) in synthetic seawater was investigated. AOM harvested from the marine diatom *Chaetoceros affinis* was mainly composed of biopolymers. These biopolymers had high fouling potential as measured by the Modified Fouling Index - Ultrafiltration (MFI-UF) and were very compressible. Filtration at higher flux exacerbated both fouling potential and compressibility of AOM. Cake/gel layers formed from amorphous AOM can be rearranged and compacted at higher filtration flux, resulting in higher pressure development.

Coagulation substantially reduced fouling potential and compressibility of AOM. At coagulant dose of 1 mg Fe(III)/L and higher, AOM fouling characteristics diminished and those of iron hydroxide precipitates prevailed, indicating that the predominant coagulation mechanism is adsorption of biopolymers on and enmeshment in iron hydroxide precipitates forming Fe-biopolymer aggregates. At low coagulant dose, fouling potential and compressibility were marginally improved as coagulation most likely resulted in partial complexation of the biopolymers and formation of colloidal Fe-biopolymer complexes. Coagulation also substantially reduced the flux dependency of AOM filtration, resulting in substantially lower MFI-UF values and linear development of pressure in filtration tests at constant flux.

At a low coagulant dose (~ 0.5 mg Fe(III)/L), inline coagulation/UF was more effective in terms of biopolymer removal than conventional coagulation followed by sedimentation and filtration (0.45 μm). However, to achieve removal rates of more than 80% of biopolymers with inline coagulation/UF, coagulant dose of ≥ 5 mg Fe(III)/L was required.

9.1.3 Minimizing coagulant dose in UF pretreatment to SWRO

The applicability of coating UF membranes with a removable layer of particles at the start of each filtration cycle for operation on algal bloom-impaired seawater was investigated. For this purpose, iron hydroxide particles were applied as coating material at the start of each filtration cycle at different equivalent dose.

Algal organic matter concentrations in the range of 0.2 - 0.7 mg C/L as biopolymers showed poor backwashability of the UF membranes. Coating with ferric hydroxide particles prepared by simple precipitation and low intensity grinding was effective in controlling non-backwashable fouling. However, relatively high equivalent dose (~ 3 - 6 mg Fe(III)/L) was applied. Coagulant dose was minimized to low levels (~ 0.3 - 0.5 mg Fe(III)/L) by reducing the particle size of the coating material. Coating with ferric hydroxide particles smaller than 1 μm - prepared by precipitation followed by high intensity grinding - resulted in stable operation of the UF membranes at low dose (≤ 0.5 mg Fe(III)/L).

The success of coating in reducing non-backwashable fouling of UF membranes is attributed to a combination of physical and chemical mechanisms. Results indicated that physical shielding may not be the only mechanism - as initially hypothesized - and (chemical) interaction between coating material and the biopolymers attached to the membrane surface also occurs in such a way that non-backwashable fouling is reduced.

9.1.4 Algal organic matter removal from SWRO feed water

(Bio)fouling of RO membranes and/or spacers is a major concern for SWRO systems, resulting in the need for frequent chemical cleaning of the membranes. Low cleaning frequency of SWRO membranes is desired to minimize the risk of damaging the membranes, downtime and quantities of chemicals to be discharged. This is in particular true for large scale seawater reverse osmosis plants which have been put in operation in the last decade. AOM biopolymers (including transparent exopolymer particles, TEP) can initiate and exacerbate biofouling in SWRO membranes. Therefore removal of AOM from SWRO feed water with effective pretreatment is required. The extent to which AOM can be removed from SWRO feed water was studied for conventional coagulation (i.e., coagulation, flocculation, sedimentation, filtration) and a coagulant free alternative (i.e., low molecular weight cut-off UF).

Coagulant dose of up to 20 mg Fe(III)/L was required for coagulation/flocculation followed by sedimentation to reduce biopolymer concentration by up to 70%. Filtration (through membranes with 0.45 μm pore size) had a significant impact on AOM removal, even at coagulant dose < 1 mg Fe(III)/L. This indicated that coagulated AOM aggregates have better filterability than settlability characteristics. This may be attributed to the low density of these aggregates and has considerable implications for the choice of clarification process in conventional pretreatment systems.

UF membranes with a nominal molecular weight cut-off of 10 kDa were applied as a coagulant free alternative for reducing fouling potential of SWRO feed water during algal bloom periods. 10 kDa membranes were capable of completely removing AOM biopolymers. UF membranes with nominal molecular weight cut-off of 150 kDa (without coagulation) were capable of reducing biopolymer concentration to approximately 200 μg C/L (~ 60% removal). In terms of hydraulic operation, 10 kDa membranes showed lower permeability recovery after backwash than 150 kDa membranes. This was attributed to the low surface porosity of the 10 kDa membranes, as determined through FE-SEM images.

9.2 General Outlook

The findings of this study suggest that coagulant consumption may be substantially reduced in UF pretreatment of SWRO by applying alternative solutions. Coagulation improved UF hydraulic performance mainly by reducing non-backwashable fouling of the membranes. This can be achieved at very low coagulant dose by coating the membranes with sub-micron particles. Further reducing particle size of the coating suspensions to the lower nanometre range is likely to be more effective in reducing the required equivalent dose and is recommended for future research.

Currently, very high coagulant dose (up to 20 mg Fe(III)/L) is needed to reduce AOM biopolymers in SWRO feed water to less than 100 μg C/L. UF membranes with nominal molecular weight cut-off of 150 kDa require lower coagulant dose to remove AOM biopolymers. Low molecular weight cut-off UF membranes are capable of completely removing AOM biopolymers in RO feed water without coagulant addition. Further improvement in material properties of these membranes is expected to realize their full potential in SWRO pretreatment.

In general terms, this study indicates the applicability of UF membranes with low molecular weight cut-off as a coagulant free future of SWRO pretreatment. Major benefits in terms of reduced environmental impacts are expected when applying membranes with low molecular weight cut-off, as the need for coagulation is eliminated while ensuring longevity of downstream SWRO membranes. Further improvements in material properties of these membranes should be directed at increasing the surface porosity of the membranes to enhance permeability recovery and ensure stable hydraulic operation.

Abbreviations

AAS	Atomic absorption spectroscopy
AOC	Assimilable organic carbon
AOM	Algal organic matter
APHA	American Public Health Association
ATP	Adenosine triphosphate
BB	Building blocks
BDOC	Biodegradable organic carbon
BOD	Biological oxygen demand
BW	Backwashable
CDOC	Chromatographable dissolved organic carbon
CEB	Chemically enhanced backwash
CIP	Cleaning in place
COC	Chromatographable organic carbon
DAF	Dissolved air flotation
DLS	Dynamic light scattering
DOC	Dissolved organic carbon
ED	Electrodialysis
EDTA	Ethylenediaminetetraacetic acid
EOM	Extracellular organic matter
EPA	Environmental protection agency
ESA	European space agency
F-EEM	Fluorescence excitation emission matrix
FE-SEM	Field emission scanning electron microscope
FTIR	Fourier transform infrared spectroscopy
GMF	Granular media filtration
GWI	Global water intelligence
HAB	Harmful algal bloom
HOC	Hydrophobic organic carbon
HS	Humic substances
ICP-MS	Inductively coupled plasma - mass spectrometry
IOM	Intracellular organic matter
IR	Infrared

LC-OCD	Liquid chromatography - organic carbon detection
LoD	Limit of detection
MED	Multi-effect distillation
MF	Microfiltration
MFI	Modified Fouling Index
MFI-UF	Modified Fouling Index - Ultrafiltration
MSC	Microbial support capacity
MSF	Multistage flash
MSF	Membrane fouling simulator
MW	Molecular weight
MWCO	Molecular weight cut-off
nBW	Non-backwashable
NDP	Net driving pressure
NF	Nanofiltration
NPOC	Non-purgeable organic carbon
NTU	Nephelometric turbidity unit
OCD	Organic carbon detector
OND	Organic nitrogen detector
PACl	Polyaluminium chloride
PES	Polyethersulphone
PSD	Pore size distribution
PVDF	Polydivinylidene difluoride
RC	Regenerated cellulose
RO	Reverse osmosis
RSF	Rapid sand filter
RU	Raman units
SDI	Silt density index
SEC	Size exclusion chromatography
SEM	Scanning electron microscope
SSW	Synthetic seawater
SUVA	Specific ultraviolet absorbance
SWRO	Seawater reverse osmosis
TEM	Transmission electron microscope
TEP	Transparent exopolymer particles
TMP	Transmembrane pressure
TOC	Total organic carbon
TSS	Total suspended solids
UF	Ultrafiltration
UV	Ultraviolet
UVD	Ultraviolet absorbance detector
WHO	World Health Organisation

Publications and awards

Peer-reviewed journals

Tabatabai, S.A.A., Schippers, J.C., Kennedy, M.D., 2014. Effect of coagulation on fouling potential and removal of algal organic matter in ultrafiltration pretreatment to seawater reverse osmosis. Water Research, *In Press*.

Schurer, R., **Tabatabai, A.**, Villacorte, L., Schippers, J.C., Kennedy, M.D., 2013. Three years operational experience with ultrafiltration as SWRO pre-treatment during algal bloom. Desalination and Water Treatment 51 (4-6), 1034-1042.

Tabatabai, S.A.A., Jin, W., Kennedy, M.D., Amy, G.L., Schippers, J.C., 2010. Examining coagulation and precoating for seawater UF/RO system pretreatment. IDA Journal (Desalination and Water Reuse) 2, 58-62.

Tabatabai, S.A.A., Gaulinger, S.I., Kennedy, M.D., Amy, G.L., Schippers J.C., 2009. Optimization of inline coagulation in integrated membrane systems: a study of $FeCl_3$, Desalination and Water Treatment 10, 121 – 127.

Tabatabai, S.A.A., Kennedy, M.D., Amy, G.L., Schippers, J.C., 2009. Optimizing inline coagulation to reduce chemical consumption in MF/UF systems, Desalination and Water Treatment 6, 94 – 101.

Conference proceedings

Tabatabai, S.A.A., Kennedy, M.D., Schippers, J.C., Amy, G.L., 2012. Coating UF membranes with iron nanoparticles to enhance operation during periods of severe algal bloom. NAMS 22nd Annual Meeting, New Orleans, June

Tabatabai, S.A.A., Hernandez Caballero, M., Hassan, A., Ghebremichael, K., Kennedy, M.D., Schippers, J.C., 2012. Low dose coagulation to enhance UF operation during algal blooms. EDS Conference Desalination for the Environment: Clean Water and Energy, Barcelona, April

Tabatabai, S.A.A., Hernandez Caballero, M., Hassan, A., Ghebremichael, K., Kennedy, M.D., Schippers, J.C., 2011. Optimizing coagulation in seawater UF/RO to reduce fouling by transparent exopolymer particles (TEPs). IDA World Congress, Perth, September

Tabatabai, S.A.A., Jin, W., Kennedy, M.D., Schippers, J.C., 2010. Pretreatment options to reduce biofouling potential in seawater reverse osmosis systems. EDS Conference EuroMed, Tel Aviv, October

Tabatabai, S.A.A., Jin, W., Kennedy, M.D., Schippers, J.C., 2010. Role of coagulants in seawater reverse osmosis pretreatment systems. 3rd Oxford Water and Membranes Research Event, September

Tabatabai, S.A.A., Jin, W., Kennedy, M.D., Amy, G.L., Schippers, J.C., 2010. (Inline) coagulation or precoating for seawater reverse osmosis systems? AMTA Conference, San Diego, July

Tabatabai, S.A.A., Jin, W., Kennedy, M.D., Schippers, J.C., 2010. (Inline) coagulation or precoating for seawater reverse osmosis systems? EDS Conference, MDIW, Trondheim, June

Tabatabai, S.A.A., Gaulinger, S.I., Kennedy, M.D., Amy, G.L., Schippers, J.C., 2008. Optimization of inline coagulation in integrated membrane Systems: a study of $FeCl_3$. EDS Conference EuroMed, Dead Sea, November

Tabatabai, S.A.A., Kennedy, M.D., Amy, G.L., Schippers, J.C., 2008. Optimizing inline coagulation to reduce chemical consumption in MF/UF systems. EDS Conference MDIW, Toulouse, October

Awards

Nominated for Gijs Oskam Award, Best MSc thesis on drinking water treatment, November 2007

Oral award for best student paper at European Desalination Society conference on Membranes in Drinking Water Production and Wastewater Treatment (MDIW) 2008, Toulouse, October 20-22

Best student paper award at American Membrane Technology Association conference "The wave of the future has arrived" 2010, San Diego, July 12-15 (presented by L.O. Villacorte)

∞

Biography

S. Assiyeh Alizadeh Tabatabai was born in Tehran, Iran on 29 August 1982. She received her B.Sc. in Civil Engineering in 2005 from the University of Engineering and Technology, Lahore, Pakistan. After graduation, she joined geotechnical consultants in Tehran and worked on feasibility studies for Tehran Metro. In 2007 she obtained her MSc in Water Supply Engineering from the Department of Water Supply and Sanitation. Her thesis entitled "Optimizing inline coagulation to reduce chemical consumption in MF/UF systems" was nominated for a Gijs Oskam award.

From 2007 - 2009, she was involved in a training and capacity building project, funded from a World Bank loan, for the water and sanitation sector in Iran. In December 2009, she started working on her PhD research at the Department of Environmental Engineering and Water Technology. Her research was part of a larger project "Zero Chemical UF/RO for Seawater Desalination" in collaboration with Norit, Evides, Vitens, RWTH and University of Twente. The project was funded by the Dutch Ministry of Economic Affairs (SenterNovem) and supported by Norit X-Flow.

Ms. Tabatabai is actively involved in several water and desalination associations and has served as European Coordinator of the International Desalination Association - Young Leaders Programme from 2011 to 2013. She continues to be on the Young Leaders Programme committee as Technical Coordinator with an active interest in university outreach focused on awareness-raising on the desalination industry at university level.

* 9 7 8 1 1 3 8 0 2 6 8 6 5 *
T - #0395 - 101024 - C34 - 240/170/14 - PB - 9781138026865 - Gloss Lamination